# THE CALENDAR FOR EVERYBODY

# THE CALENDAR
## *For Everybody*
### *by*
### ELISABETH ACHELIS

## G.P. PUTNAM'S SONS
### NEW YORK

Republished by Omnigraphics • Penobscot Building • Detroit • 1990

Library of Congress Cataloging-in-Publication Data

Achelis, Elisabeth, 1880–
    The calendar for everybody / by Elisabeth Achelis.
        p.   cm.
    Reprint. Originally published: New York : Putnam, 1943.
    ISBN 1-55888-849-7 (lib. bdg. : alk. paper)
    1. Calendar.   2. Calendars.   I. Title.
CE11.A34   1990
529′.3—dc20                                                89-63208
                                                                CIP


⊗ This book is printed on acid-free paper meeting the ANSI Z39.48 Standard. The infinity symbol that appears above indicates that the paper in this
book meets that standard.

Printed in the United States of America.

This book
is dedicated to
the hosts of friends, advocates, workers
and calendar committees throughout the world
in grateful recognition
of their understanding and unswerving support
for the new and better time-plan
THE WORLD CALENDAR

# CONTENTS

vii

viii          *Contents*

# FOREWORD

TIME, WHETHER WE REALIZE it or not, is so closely inter-woven with our every thought and action that we all per-form our daily and various tasks in complete compliance with it. Unswerving and relentless, Time measures the span of our lives and records individual, communal, national and world affairs.

Is the calendar—the recorder of Time—the best instru-ment we could have? Does it serve us well in every way? Or is it a willful taskmaster by which we are daily and hourly tried?

This narrative of the calendar in its beginning and devel-opment and in its characteristics good and bad, introduces the calendar as a friend. It becomes a human document and arouses in us a warmer, more intimate feeling, thereby mak-ing it easier to correct its numerous shortcomings. No longer need we put up with these because our forebears did. No longer need we think of the calendar as inviolate and im-possible to change, or that present conditions make improve-ments untimely.

Worn out and faulty machines and instruments are not

tolerated but discarded and perfected, and we shall see that our present civil calendar, too, is outmoded and requires improvement. A new and better time-plan is as vital to war conditions, to postwar issues as to days of peace. Perhaps even more so because the best machines, the best instruments and the best plans spell progress and success. The calendar should grow up and develop as generations do, and therefore should have the right to be changed and improved, better to serve every one of us.

The following chapters tell of little-known facts of our present outmoded calendar and the advantages to be had from the new, simplified and steadfast calendar for everybody's use.

ELISABETH ACHELIS

*June, 1943.*

# THE CALENDAR FOR EVERYBODY

*Time and the Calendar*
*In their ceaseless rounds upon rounds*
*Are inseparable companions.*

# 1

# COMPASS, CLOCK AND THE CALENDAR

WEBSTER'S INTERNATIONAL DICTIONARY defines the calendar as:

"a system of fixing the beginning, length, and division of the civil year; an orderly arrangement of divisions of time as years, months, weeks, and days, adapted to the purposes of civil life."

The civil calendar, our measurer of Time, is actually the compass for our days and years by which we chart our lives—the many activities and events that occur in our personal, communal, national and international affairs. It is similar to the compass that directs ships in their voyagings across the seas, and the clock as it marks our days in its charted course of twenty-four hours. The calendar, the clock and the compass are indispensable instruments in our civil life.

Can you imagine what life would be without a calendar that tells of intervals and associations of events? Would we not be laboring in a hopeless labyrinth of unrelated events? Every act would be one of isolation without focus, direction or meaning. History could no longer be our mentor admonishing us of past errors and inspiring us to carry on in the spirit of the great men and women of the past. Our days would be a sorry mass of confusion and we would be lost in a maze of helpless bewilderment. We would be like rudderless ships drifting without purpose. The civil calendar is one of the most important organized institutions we have today. A date and a year immediately associate for us the past and the present.

There are certain outstanding events that have left their indelible imprint on civilization.

The birth of Christ, 1 A.D., observed December 25, links Palestine with Roman domination and the birth of a new religion. In 321 A.D., Constantine the Great introduced into the Julian calendar the seven-day week with its Sunday, the day of rest, recognizing thereby Christianity over paganism. The invasion and conquest of England by the Normans, 1066 A.D., freed her from past isolation whereby she became more closely related and united with the European continent. The Magna Carta, 1215 A.D., laid the foundation for personal freedom; and the discovery of the Western World by Christopher Columbus, October 12, 1492, came at the time of the conquest of Granada by Spain which so enriched her exchequer that she could finance Columbus in his daring venture.

The Gregorian calendar by Pope Gregory XIII, in 1582, was enacted during the days of the great Reformation; the American Revolution founded a new form of government, as proclaimed in the Declaration of Independence, July 4, 1776, while France was unfriendly to England and on the threshold of her own revolution; and the civil war of the United States begun April 12, 1861, established freedom of man from slavery in the years when the ill-fated Maximilian expedition to Mexico was undertaken to form a kingdom. These and hundreds of other events reveal the close association one event has upon another as disclosed through the medium of the civil calendar.

The compass that guides ships at sea and airplanes in the air is a master instrument of perfection. Its focal point is the magnetic north, that place on the earth near the North Pole toward whose mysterious power the magnetic needle invariably points. The compass is a reliable instrument.

The clock is equally dependable. Its central point is the midnight hour 12 P.M. (known in some parts of the world as 24 o'clock) that marks the end of one day and immediately on reaching the midnight hour announces the split second of the new day. The change is instant. The clock is our unfailing guide.

In contrast, the civil calendar has *no focal point*. It is true that the calendar always begins each new year on January first, but it fails utterly to mark the weekday. And herein lies a strange psychological mystery. Whereas man is meticulous in reckoning and arranging his everyday life by the weekday, yet in his annual recording by years, months and dates, he completely ignores, by some strange quirk, the day. But has not the day its particular characteristic as has the month?

Because of this strange omission we can never link the weekday with the month date in our civil calendar unless we resort to a "perpetual" calendar that few people possess, or do arduous research in libraries, or study and read historical books, or perform mathematical gymnastics.

For instance, it is of the utmost value to know that Martin Luther posted his famous thesis on the door of the Castle Church at Wittenberg on October 31, 1517, a Saturday, which marked the beginning of the Reformation. He se-lected that particular day so that church-goers could see and read it the following day without delay. The day, as well as the date, was important. So it is also well to realize that Christopher Columbus first sighted the New World on Friday, October 12, 1492, and landed Saturday, the 13th.

Likewise, we faithfully observe the Fourth of July, but the fact that the Declaration of Independence was adopted by Congress on a Thursday is little known. The first shot fired at Fort Sumter, which started the Civil War, was on April 12, 1861, a Friday, and the tragic shot that killed Lincoln was strangely enough also fired on a Friday—Good Friday— April 14, 1865. The date, April 6, 1917, on which the United States declared war against Germany in the first World War, was also a Friday, another Good Friday. Are these week days without significance? It is interesting to observe that the North Pole was discovered by Commander Robert E. Peary, on April 6, 1909, a Tuesday, another indication of the irregularity of the calendar that one April 6 was a Fri-day, the other a Tuesday.

These various illustrations show the significance the day has with its date, the focal point which the present calendar lacks.

Then, too, the fact that Hitler invaded countries during week-ends will be practically lost to historians because days and dates do not agree from one year to another. Dating and recording can never be complete, never comparable. The calendar is a shifting instrument; it certainly is not the very best for faithful recording. It falls far short of the compass and the clock.

To achieve for the civil calendar the same steadiness and reliability that the compass and the clock give, and to order its arrangement on an equitable basis are the objectives of an improved calendar. The calendar, like the compass and the clock, must be the very best under all conditions.

To ignore the revision of the calendar just because it is a prosaic, mathematical and familiar instrument, to adhere to the false premise that it must wait for a more propitious time, and to ignore its defects, are not justifiable reasons for delaying the change. The confusion and unreliability of the calendar are constantly carried over into our civil life. Realizing this, should we not revise the calendar as quickly as possible? Does it not become a privilege, nay a duty, for this generation, especially during these tense and disturbed war years, to install a calendar that is really "an orderly arrangement" and the best "adapted to the purposes of civil life," whereby it may better serve us?

Improvements and reforms are more often effected during distressful and difficult days than when people live in ease and comfort as the following pertinent inscription on the cornerstone of a chapel in Leicestershire, England, shows:

> "In the year 1653
> When all things sacred
> were throughout ye nation
> either demolished or profaned
> Sir Robert Shirley, Baronet,
> builded this church;
> whose singular praise it was
> *to have done the best things*
> *in the worst times*
> *and to have hoped them*
> *in the most calamitous.*"[1]

[1] Italics mine, E. A.

In this world-struggle to let go of the yesterdays and to accept the tomorrows, inevitable changes are taking place in our todays in every thing, in every way, in every nation and in every individual.

We are experiencing and gaining a new understanding of the dignity of man, new forms of government (national and international), and new international relationships. New discoveries in the sciences, new interpretations of religion, new concepts of economics, a new outlook on education, new manifestations in art and culture, new means of transportation and communication, and a new system of calendar, are all in the making.

Time, as we know it, came with the creation of earth. The very words: "In the beginning," God gave to man his human time-sense. With the steady progression of Time and man's increasing knowledge and development, his sense of Time as expressed in the civil calendar also changed. Of these changes there have been many in the past and today it is logical and evident where everything is being made new that a new calendar also is inevitable. For a new and better world cannot be built on a calendar with its faulty pattern of yesterday.

Shall we not, then, with open and free minds adopt a better civil calendar which with many other new things is knocking at the door of our world, seeking acceptance?

# 2

# EARTH'S TIME

On a warm summer day, sitting under the pine trees beside a rambling bungalow at Grand Isle, Vermont, on the border of Lake Champlain, I was lazily watching the rhythmic movement of the lake. Without sound the water moved in slow undulating wavelets. It was like the quiet breathing of a sleeping child. And as I watched, I felt this movement to be Earth's breath. Everywhere and in everything I seemed to be conscious of this quiet movement, this continuous breathing, that is taking place in trees, grass and flowers, in rock and soil, in mountains and valleys, in rivers, lakes and oceans, in country and city. Earth is constantly breathing, and her breath is in every living thing that inhabits Earth, whether under the soil, in the waters, on the surface, or in the air.

I then began to ponder how many breaths Earth had taken since her formation and that with every breath she took there was life, there was movement. There was the past breath, the breath of the moment, the breath to come. That which was present instantly became the past and that which was future had become as quickly the present. There was the movement, too, of Earth's daily rotation on her axis that gives her the day and the night, and the annual revolution around the sun that gives her the seasons and the year.

This I conceived as Time. Time is Earth's breath. Time is the steady rhythmic movement of Earth. It came with Earth and will sustain her while Earth lasts. No one knows when Earth took her first breath and Time came, and no one knows if she will exhale her last breath and Time will cease. How defineless and limitless is Time! How mysterious the influence that it exerts in the life of every one of us!

With this thought Time became tremendously significant to me and took on new meaning. How ageless was Time! How many millions, billions, nay trillions and quadrillions of breaths Earth must have taken to have attained our today! I understood better than ever before "that one day is with the Lord as a thousand years, and a thousand years as one day." [2]

From a living formless mass through untold aeons of years, Earth has evolved from dark, misty vapors to the light, clear globe we know as Earth. How she must have groped and crawled and stumbled in her various stages of formation! How she must have been experimented upon during these long eras of development!

Astronomers and geologists compute the Earth as hundreds of millions of years old and it is tens of millions of years since life appeared. Man, they tell us, is but a few million years old and this by comparison makes him a mere child. From the gaseous to the fluid, from swamp, rock and plant, to marine and amphibious life, the reptile and the mammal, through the glacial, bronze and iron ages, down to our present machine age, continuous life has flowed,

[2] 2 *Peter,* 3:8.

numerous transitions have taken place, untold myriads of
Earth's breaths have been taken.

The majestic and symbolic story of creation to me became enormously enriched with new significance. I could
not but contemplate the marvelous manifestations of Earth,
when from a formless mass she became a sphere, and
evolved the various forms of life. Surely this was not accomplished suddenly, but through incalculable numbers of
Earth's breaths, which we shall now call TIME.

Through the wise guidance and loving care of the Creator
a beautiful pattern was being formed. The separation of
heaven and earth was followed by the appearance of land,
and by means of the celestial stars, moon and sun, and the
days, seasons and years, Earth's movement could be measured. The sun gave her the light of day, the moon bright-
ened the dark of night.

Aeons must have passed into the vast stages of development, all in preparation for human life. With the coming
of man, made in the image and likeness of the Creator, he
became the supreme being to inhabit and to dominate the
Earth, and the long story of creation in its visible form was
complete. But the cycle of man's growth and development
within boundless Time, Earth's breathing, is not complete.
Surely there lie unknown centuries ahead of him wherein to
grow and to expand until, as we are told, he shall have
attained his divine perfection and is one with his Creator.
This, I perceived as the inspiring plan of time, of creation,
and of the promise by which man is being led onward
and upward to a fuller and more complete understanding
of himself, of life, and of God. And in this, Time is closely
related and plays an important part.

# 3

# EARLIEST TIME

M AN, COMING ON EARTH, must have been instantly aware
of the day of light and the night of dark, which in their
ceaseless rounds upon rounds never varied. It was an orderly
rotation upon which he could entirely depend. This was
surely a blessed boon to him in the many vicissitudes that
confronted him during his early uncertain and precarious
days on the planet Earth.

He used his days in search of food to sustain him, his
nights to rest from his labors. These he counted as so many
light and so many dark periods. Scientists and historians tell
us that early man gathered sticks and pebbles and cut notches
in trees and poles to tell the numbers of these continuing
periods of light and dark. H. G. Wells in his *Outline of
History* writes:

"And Neolithic man was counting, and falling under the spell of
numbers. There are savage languages that have no word for any
number above five. Some peoples cannot go above two."

The ten fingers and ten toes were also convenient digits
to count by and it is said that the score of twenty had its
origin with the twice told ten.

The earliest reckoning of Time by days and nights must have lasted centuries since man is a slowly evolving creature. Man gradually became conscious of the moon as a time-recorder, which shone so bright and full some nights and on others became dimmed and entirely disappeared. He thus progressed to a longer period of Time, the moon-periods, as being more convenient and easier to remember. He began to count the new moons to reckon the birth of his child, for festival days, and to celebrate certain events in his personal and tribal life. The regular progression of new, waxing, full and waning moons was man's celestial clock and became his first calendar. By the moon the years were reckoned and generations of men were recorded.

There was no unified system of counting Time by the moon, however, since the different peoples that inhabited earth lived in various localities and had their own manner of counting. Some began their months and years with the new moon and others with the full moon. This made little difference in these early days of man's earth-life as these groups were more or less isolated. Tribal groups had their own distinct communities and customs; climatic and geographical differences, too, influenced them variously. It is all the more significant to observe that whether man lived in China, India, Europe, Chaldea or Babylon, the moon calendar appears everywhere to have been the generally approved method of counting Time, although it differed in detail.

Man always has been a worshiper. That which he cannot comprehend is the irresistible magnetic needle. He ever looks to some being higher than himself for protection, for guidance and for propitiation. Dimly he seemed to remember from whence he came, after whose image he was made, and the divine promise of his ultimate return. Wonderingly and worshipfully he observed the mysterious moon so far off in the heavens. Eagerly he awaited the first appearance of the thin crescent which caused in him a spirit of rejoicing and marked a special day of recording. Just as eagerly he watched for the culmination of the full moon when the dark and fearsome nights became bright and seeing, marking another significant day in his life, only to see the brightness disappear again into the eerie blackness of the night. These regular phases of the moon were guideposts in his regular days that cast an enthralling spell on him as they fascinate us today. The goddess "Diana, huntress chaste and fair," still casts her sway over many of us.

The early moon calendars consisted of 12 months, each of 29 or 30 days. Since it takes the moon about 29½ days to complete a cycle from one new moon to the next, and to complete a year, it was necessary to resort to alternating numbers of days in the months. Thus it was that the moon gave to man the name "moonth—month." The lunar year has 354 days with an occasional 13th month inserted. In *The Romance of the Calendar,* P. W. Wilson reminds us that in primitive times the year was a vague and varied unit.

It is reasonable to assume that man had been living and counting by moon calendars for many centuries before he advanced another step in his development and became aware of and was willing to accept a newer and better time-measurer; one that was more reliable than the changeable and inconstant moon.

From hunting and roaming for food he had gradually become a herdsman and an agriculturalist. He began to tend his animals and to cultivate the soil for nourishment. And with the need for tilling, sowing, growing and harvesting of crops he became keenly aware of nature's regular and rhythmic seasons. Studying the stars and the sun he learned of the orderly progression of the seasons and that each of these was accompanied by its own particular weather and climate. This knowledge was vital because it enabled him to sow and to increase his crops with greater assurance of success and better to protect himself and his animals which were rapidly being domesticated.

It takes little imagination to realize the far-reaching influence this had upon man's entire life and his many activities. And yet this new knowledge came gradually, first to the few, then to the many. The overcoming of the old idea that the moon directs Time with all that it implies, and the acceptance of the new, that the sun influences the seasons and directs the years and therefore is a more dependable time-recorder, was not received without opposition and protest. New things are not readily accepted. The old, that which is habitual, exerts a tenacious hold on man. And man, being a creature of habit, and leaders, fearful of losing their strongly built power, resist change. Yet it is only with change that man ever grows, develops and reaches out to better things and better conditions. To stand still is but slow stagnation certain to result in disintegration, which may take centuries and then again a comparatively few years.

This resistance to new and better things is as apparent in calendar history as in every other field. To make things better takes vision, courage, determination and faith, and leaders of calendar reform have been richly endowed with these qualities.

# EGYPTIAN CALENDAR

## EACH YEAR THE SAME

| FIRST | | | SECOND | | | THIRD | | | FOURTH | | |
|---|---|---|---|---|---|---|---|---|---|---|---|
| 1 | 11 | 21 | 1 | 11 | 21 | 1 | 11 | 21 | 1 | 11 | 21 |
| 2 | 12 | 22 | 2 | 12 | 22 | 2 | 12 | 22 | 2 | 12 | 22 |
| 3 | 13 | 23 | 3 | 13 | 23 | 3 | 13 | 23 | 3 | 13 | 23 |
| 4 | 14 | 24 | 4 | 14 | 24 | 4 | 14 | 24 | 4 | 14 | 24 |
| 5 | 15 | 25 | 5 | 15 | 25 | 5 | 15 | 25 | 5 | 15 | 25 |
| 6 | 16 | 26 | 6 | 16 | 26 | 6 | 16 | 26 | 6 | 16 | 26 |
| 7 | 17 | 27 | 7 | 17 | 27 | 7 | 17 | 27 | 7 | 17 | 27 |
| 8 | 18 | 28 | 8 | 18 | 28 | 8 | 18 | 28 | 8 | 18 | 28 |
| 9 | 19 | 29 | 9 | 19 | 29 | 9 | 19 | 29 | 9 | 19 | 29 |
| 10 | 20 | 30 | 10 | 20 | 30 | 10 | 20 | 30 | 10 | 20 | 30 |

| FIFTH | | | SIXTH | | | SEVENTH | | | EIGHTH | | |
|---|---|---|---|---|---|---|---|---|---|---|---|
| 1 | 11 | 21 | 1 | 11 | 21 | 1 | 11 | 21 | 1 | 11 | 21 |
| 2 | 12 | 22 | 2 | 12 | 22 | 2 | 12 | 22 | 2 | 12 | 22 |
| 3 | 13 | 23 | 3 | 13 | 23 | 3 | 13 | 23 | 3 | 13 | 23 |
| 4 | 14 | 24 | 4 | 14 | 24 | 4 | 14 | 24 | 4 | 14 | 24 |
| 5 | 15 | 25 | 5 | 15 | 25 | 5 | 15 | 25 | 5 | 15 | 25 |
| 6 | 16 | 26 | 6 | 16 | 26 | 6 | 16 | 26 | 6 | 16 | 26 |
| 7 | 17 | 27 | 7 | 17 | 27 | 7 | 17 | 27 | 7 | 17 | 27 |
| 8 | 18 | 28 | 8 | 18 | 28 | 8 | 18 | 28 | 8 | 18 | 28 |
| 9 | 19 | 29 | 9 | 19 | 29 | 9 | 19 | 29 | 9 | 19 | 29 |
| 10 | 20 | 30 | 10 | 20 | 30 | 10 | 20 | 30 | 10 | 20 | 30 |

| NINTH | | | TENTH | | | ELEVENTH | | | TWELFTH | | |
|---|---|---|---|---|---|---|---|---|---|---|---|
| 1 | 11 | 21 | 1 | 11 | 21 | 1 | 11 | 21 | 1 | 11 | 21 |
| 2 | 12 | 22 | 2 | 12 | 22 | 2 | 12 | 22 | 2 | 12 | 22 |
| 3 | 13 | 23 | 3 | 13 | 23 | 3 | 13 | 23 | 3 | 13 | 23 |
| 4 | 14 | 24 | 4 | 14 | 24 | 4 | 14 | 24 | 4 | 14 | 24 |
| 5 | 15 | 25 | 5 | 15 | 25 | 5 | 15 | 25 | 5 | 15 | 25 |
| 6 | 16 | 26 | 6 | 16 | 26 | 6 | 16 | 26 | 6 | 16 | 26 |
| 7 | 17 | 27 | 7 | 17 | 27 | 7 | 17 | 27 | 7 | 17 | 27 |
| 8 | 18 | 28 | 8 | 18 | 28 | 8 | 18 | 28 | 8 | 18 | 28 |
| 9 | 19 | 29 | 9 | 19 | 29 | 9 | 19 | 29 | 9 | 19 | 29 |
| 10 | 20 | 30 | 10 | 20 | 30 | 10 | 20 | 30 | 10 | 20 | 30 |

*Five additional days at end of year— 1 2 3 4 5*

EGYPT HAD THREE EQUAL SEASONS OF FOUR MONTHS EACH:
INUNDATION, SOWING, HARVESTING.

# 4

# EGYPTIAN SUN CALENDAR

To the Egyptians, the Nile was the giver of life. Without this river the land would have remained barren, the peoples undernourished and impoverished. The Nile was the life stream around which their cultural, social, agricultural, physical, national and religious life was built.

It was of the utmost importance, then, to have the exact knowledge of the times when the Nile overflowed its banks leaving a rich deposit which fertilized the soil, when it receded to its regular course to allow for the tilling and sowing of the land, and the growing and harvesting of crops. Astronomers and priests scanned the sky, studied the stars, observed the movement of the moon, and measured the shadows cast by the pyramids to ascertain the sun's position, to increase their knowledge whereby they could discover the great mystery of their river Nile. Their quest for knowledge was unceasing. Thus they erected the remarkable and little understood Great Pyramid at Gizeh—a mighty structure of a vast mathematical concept. It aided the Egyptians in their study of the heavens and was used as a sacred temple and tomb; a symbol that combined in one stupendous edifice their science, their religion, their philosophy and their thought. It stands as an enduring master achievement for all ages.

With painstaking and determined effort lasting many years, the Egyptians discovered that after many days of invisibility, the first reappearance in the eastern horizon of their most brilliant star Sirius (also called Sothis) came just before the rising of the sun, known as the "heliacal" rising of Sirius, which they thought occurred regularly after an interval of 360 days. This successive heliacal rising of Sirius, the Dog Star, gave the Egyptians their unique time-period of 360 days, which many believe was the origin of the 360 degree circle. This complete cycle of days became their year. It is easily understood that the regular reappearance of Sirius was a most notable celestial and annual event in ancient days, and marked the beginning of every new year. It was the pivot for their astronomical reckonings, and historians and scientists tell us that the new year's day in antiquity was July 19.

Then something very amazing was discovered which was to solve the quest of the Egyptians. They learned that the annual reappearance of the "calendar star" Sirius came almost simultaneously with the rising of the Nile, so that these two—the star Sirius and the river Nile—became closely associated in the life of the people. This led to the fuller knowledge that the four-month season of inundation, in this ancient nature calendar, was followed by a four-month season of tilling and sowing. Then came the harvest season of four months. These three seasons, each of four months, were then combined into one whole unit and became a year. This was the transitional star or sidereal calendar that came between the earlier moon and the later sun calendars. It was a calendar of 12 months, each month having 30 days.

The stellar calendar was a decided step forward. A tremendous innovation from the past, it proved more reliable and stable not only for astronomical, pastoral and religious purposes but for the community life of the people as well. It commanded attention and could not be ignored.

However, the change was not complete; it was but a forerunner of a still better calendar. With further intensive observations it was learned that the seasons were not dependent so much on the calendar star Sirius as upon the direct influence and location of the sun. The annual cycle from one fixed period to another identical period, such as one spring equinox to another, took 365 days, not 360. It was not sufficient to begin every year with the rising of Sirius, or with the annual overflowing of the Nile. The regular and recurrent appearances and duration of all the seasons were important. It was thus in 4236 B.C.,[3] as determined by Borchardt according to the late Professor James Henry Breasted, that a sun calendar of 365 days based on the regular seasons supplanted the transitional stellar calendar.

This new seasonal sun calendar of 365 days retained the twelve months, each of 30 days, these being subdivided into three decades, which today we would call three weeks of ten days each. The remaining five days were added at the end of every year, considered as feast days, and placed under the control of the priesthood. The Egyptian months were known by numbers only, as the Quakers today name the

[3] Early in 1940 this date was challenged by Jotham Johnson, professor of Archaeology at the University of Pittsburgh, who stated his belief that the beginning of the Egyptian sun calendar was June 18, 3251 B.C.

months by numbers. All the previous moon associations with the month were completely discarded. The new sun year of 12 months became the longest time-unit known to these ancient people and was recorded according to the reigns of chiefs or length of dynasties.

The great Egyptologist, James Henry Breasted, said of the date 4236 B.C.:

"This date, near the middle of the forty-third century B.C., is not only the earliest fixed date in history, but also the earliest date in the intellectual history of mankind."[4]

The Egyptians were the first to recognize the social need of a civil calendar, and he further wrote:

"The calendar was thus the beginning of a great movement in human life which carried over the thought of man from the world of nature to the world of human life."[4]

And science, too, gives remarkable tribute to the calendar as an outstanding intellectual event.

"The recognition of the passage of time now became a primary necessity of social life. In learning to record the passage of time man learned to measure things. He learned to keep account of past events. He made structures on a much vaster scale than any which he employed for purely domestic use. The arts of writing, architecture, numbering, and in particular geometry, which was the offspring of star lore and shadow reckoning, were all by-products of man's first organized achievement, the construction of the calendar....

"Science began when man started to plan ahead for the seasons, because planning ahead for the seasons demanded an organized

[4] *Time and Its Mysteries,* Series I, pp. 72, 73. New York University Press, 1936.

body of continuous observations and a permanent record of their recurrence." [5]

The calendar, man's first organized achievement, no longer belonged to astronomical, agronomical and religious thoughts and purposes alone, it had now expanded to the civil and social life as well.

This new sun calendar was under the protection of rulers who governed, scientists who observed celestial bodies and made their mathematical calculations, and priests who presided over religious feast and fast days. These three groups combined zealously to guard it from alien knowledge since with this improved calendar, crops were greatly multiplied whereby the nation was better nourished than the neighboring countries. It marked the beginning of a civilization that prospered and grew strong, making Egypt one of the most intellectual and influential nations of ancient days.

One cannot but conjecture who gave this remarkable sun calendar to the Egyptians for their continuous reckonings and recordings. Was he a ruler, a scientist, a priest, or was he one of a group of wise men? Was he perhaps Akhenaton, famous Pharaoh of ancient Egypt, the great lover of light, of the sun, the first humanitarian and internationalist to have expressed his belief in a single God and the brotherhood of man, who might have perpetuated the sun calendar for his people? Even though a predecessor might have initiated the sun calendar, in compliance with the earlier date mentioned by Professor Breasted, it is more than probable that Akhenaton, reflecting upon the need of his people, gave it an enduring place in the national life and for all posterity.

[5] Hogben, Lancelot, *Science for the Citizen*, pp. 5, 6. Alfred A. Knopf, New York, 1938.

After the passing of forty centuries, twice as long as the Christian era, the ever watchful and progressive Egyptians discovered that a miscalculation had been made as to the actual length of the sun year. Instead of 365 days, the year was almost a quarter of a day or six hours longer. To correct this, Ptolemy III (Eurgetes I), in 238 B.C., issued a most interesting decree. This ancient command, cut in hieroglyphics and in Greek on the seven-foot chiselled surface of the Tanis or Canopus stone was discovered in Egypt by Karl Richard Lepsius in 1886. There is a cast of the stone in the National Museum at Washington, D. C., now crated for the duration of the war. The bi-lingual decree,[6] translated by E. A. Wallis Budge, keeper of the Egyptian and Assyrian Antiquities in the British Museum, reads:

"And in order that it may happen that that which hath been decreed to be done at each season of the year may be done in accordance with the position which the heavens have with reference to the things which have to be performed at the present time, so that occasion may not be given, and the case may not arise, that some of the festivals which are celebrated in EGYPT in the winter should come to be observed in the summer, in consequence of the rising of SIRIUS advancing one day every four years, and on the other hand, some of the festivals which are at the present time celebrated in the summer should come in the future to be celebrated in the winter, a thing which actually happened in the times which are past, and would happen at the present time if the year consisted of 360 days and the five days, according to the directions for adding the additional days which have been observed: from this time onwards one day, as festival

---

[6] Budge, E. A. Wallis, *The Decree of Canopus,* Volume III, pp. 120-121. Kegan Paul, Trench, Trübner & Co., Ltd., London, 1904.

f the Good-doing Gods, shall be added every four years to the
ive additional days which come before the New Year, so that
t may happen that every man shall know that the small amount
of time] which was lacking in the arrangement of the seasons,
nd of the year, and in the things which passed as laws [or, prin-
iple] for the knowledge of their movements, hath been cor-
ected, and that it hath been supplied by the Good-doing Gods."

Mr. Budge gives the further information, "How long the
eform was carried out at Canopus cannot be said, but it
ertainly does not appear to have been adopted generally
hroughout Egypt." [7]

This measure met with defeat because the people were
oath to change and the priesthood was unwilling to relin-
quish its prerogative and to accept an occasional extra day
by rule in the calendar. Thoughts and customs had crys-
allized into tradition and the wise suggestion "supplied by
he Good-doing Gods" was rejected. It was Egypt, how-
ver, that first used the sun calendar, and although she
onceived the leap-year day as well, was prevented from
pplying it. We shall see that it was this Egyptian decree
which Caesar made available by introducing a regular leap-
ear day into his calendar revision. The sun calendar of the
Egyptians based on scientific accuracy became the veritable
warp and woof on which future calendar improvements
vere woven.

[7] Budge, E. A. Wallis, *The Decree of Canopus,* Volume III, p. 13.

# JULIAN CALENDAR

## EACH YEAR THE SAME

| Days of the Month | Martius Maius Quintilis October | Januarius Sextilis December | Aprilis Junius September November | Februarius |
|---|---|---|---|---|
| 1 | Calendae. | Calendae. | Calendae. | Calendae. |
| 2 | 6 | 4 | 4 | 4 |
| 3 | 5 | 3 | 3 | 3 |
| 4 | 4 | Prid. Nonas. | Prid. Nonas. | Prid. Nonas |
| 5 | 3 | Nonae. | Nonae. | Nonae |
| 6 | Prid. Nonas. | 8 | 8 | 8 |
| 7 | Nonae | 7 | 7 | 7 |
| 8 | 8 | 6 | 6 | 6 |
| 9 | 7 | 5 | 5 | 5 |
| 10 | 6 | 4 | 4 | 4 |
| 11 | 5 | 3 | 3 | 3 |
| 12 | 4 | Prid. Idus. | Prid. Idus. | Prid Idus |
| 13 | 3 | Idus. | Idus. | Idus. |
| 14 | Prid. Idus. | 19 | 18 | 16 |
| 15 | Idus. | 18 | 17 | 15 |
| 16 | 17 | 17 | 16 | 14 |
| 17 | 16 | 16 | 15 | 13 |
| 18 | 15 | 15 | 14 | 12 |
| 19 | 14 | 14 | 13 | 11 |
| 20 | 13 | 13 | 12 | 10 |
| 21 | 12 | 12 | 11 | 9 |
| 22 | 11 | 11 | 10 | 8 |
| 23 | 10 | 10 | 9 | 7 |
| 24 | 9 | 9 | 8 | 6 |
| 25 | 8 | 8 | 7 | 5 |
| 26 | 7 | 7 | 6 | 4 |
| 27 | 6 | 6 | 5 | 3 |
| 28 | 5 | 5 | 4 | Prid. Cal. Mart. |
| 29 | 4 | 4 | 3 | |
| 30 | 3 | 3 | Prid. Calen. | |
| 31 | Prid.Calen. | Prid.Calen. | | |
| DAYS | 31 | 31 | 30 | 28 - 29 |

# 5

# THE JULIAN CALENDAR

O F ANCIENT CALENDARS used in the Eastern Hemisphere none was as complicated as the Roman. The earliest was a moon calendar of 10 months that sprang from the foundation of Rome by the legendary twins, Romulus and Remus, in the year 1 A.U.C. which means *Anno Urbis Conditae* or *Ab Urbe Condita,* the year of the city's founding. Under our system of chronology we would say 753 years before the Christian era. Thus we note that it is about thirty-five centuries younger than the sun calendar of the Egyptians. Although Egypt bordered the southern coast of the Mediterranean and Italy the northern, so that geographically they were near neighbors, yet in culture, thought and religion they were far apart.

The successor of Romulus was the peace-loving dictator, Numa Pompilius, who, we are told, ruled over Rome for forty-three years—715-672 B.C. It was this ruler who increased the moon calendar from a previous 10 months to one of 12 months. He placed the two new months, January, at the beginning of the year before March, and February, at the end of the year after December. This arrangement lasted until 452 B.C. when the Decemvirs, a body of ten Roman magistrates who wielded great power, changed the order by placing February immediately after January.[8]

Numa Pompilius, having arranged the calendar on a 12-month basis, unlike the Egyptians who merely numbered their months, gave to each month a name. With but two exceptions the names, notwithstanding their antiquity, have persisted through the centuries and are the same that we know today. Although these are familiar, their meanings are less known and thus it is interesting to mention them here.

*January* came from Janus, the two-faced Roman god, protector of doors and gateways, patron of the beginning of all undertakings.

*February* came from februare, meaning to purify, the month for appeasing the gods and atonement for wrongs.

*March* came from Martius, the Roman god of war. It was the first month in the ancient Roman calendar.

*April* came from aperire, meaning to open, the month when earth opens to produce new blossoms; spring growth.

*May* came from the Latin Maius, third month of the early Roman year named from the Greek Maia, eldest of the Pleiades, one of the wives of Zeus.

*June* came from Juno, chief Roman goddess, protectress of women.

*July* named for Julius Caesar, replaced the old name Quintilis, meaning five.

*August* named for Augustus Caesar, replaced the old name Sextilis, meaning six.

*September* means seventh, but is the ninth month.

*October* means eighth, but is the tenth month.

*November* means ninth, but is the eleventh month.

*December* means tenth, but is the twelfth month.

[8] Black, F. A., *The Calendar and Its Reform,* p. 15. Gall & Inglis, London, 1932.

In addition, Numa Pompilius realized that the moon calendar of 12 months of alternating 29 and 30 days was too
short for the seasonal year so that it became imperative to
add an occasional 13th month whereby the calendar caught
up with the seasons. The occasional month *Mercedonius*
was inserted toward the end of February, and this method
is scientifically known as an "intercalation."

Realizing that the moon-sun calendar called for occasional
adjustments, Numa Pompilius, having established the College of Pontiffs, gave to it, together with other duties, the
authority to deal with the calendar. From the general belief
of the Egyptians to share the responsibility of the calendar
among ruler, scientist and priest, and the recognition of its
civil and social nature, the Roman calendar now became
restricted to a group of priests for its welfare and progress.
This new college had as its head a Pontifex Maximus with
four pontiffs to assist him. To them was left the responsibility of intercalation and to make such changes as might
be necessary from time to time.

According to the dictionary the word "pontifex" is composed of the noun "pons" a bridge, and the verb "facere" to
make. Pontiffs were to be bridge builders. In the case of the
calendar they were to insert the occasional 13th month Mercedonius as they believed fitting, and to "bridge" days of
worship with "time and the seasons." A tremendous responsibility was placed in their hands which, alas, was misused
and became corrupt. The intercalated month, Mercedonius,
became a pawn inserted and withheld at will. Politics and
personal ambitions had entered so that with succeeding centuries the Roman calendar was reduced to a mass of con-
fusions and contradictions. It is said that 150 days in the

calendar were illegal for business transactions. And to increase the troubles, the calendar was not only jealously guarded by priests, aided and abetted by magistrates, but was kept out of reach of the public as well.

But retaliation always follows usurpation and tyranny. The calendar, so universal in purpose, belongs to everybody and cannot for long be withheld from the people. To Cneius Flavius, son of an unknown slave and thus of humble origin, belongs the gratitude of civilization for having freed the calendar from the domination of a privileged group, in this instance, priests and magistrates.

The calendar emancipator, Flavius, resented the custom of paying tithes in order to learn on what calendar date such and such a legal act was permitted. Through ways and means, unknown to us, he succeeded in obtaining a copy of this precious calendar with its prohibited days and dates, and publicly listed them on tablets hung about the Forum. This unprecedented but highly courageous act, about 304 B.C., aroused immediate public attention and approval. The priests and magistrates were indignant at losing their former prerogatives by this flagrant act of publicity. They tried all kinds of subterfuges and devised new codes, but all to no avail. The deed was done. The calendar had blazoned forth for all to see. It was the calendar's emancipation. It was the first step toward freedom and progress for both the calendar and its users. Flavius received rich rewards from the people for he was successfully elected to office of triumvir and tribune, notwithstanding his humble origin. The Flavian act paved the way for the later Julian reform.

Before we mention this, however, another significant

change in the calendar should be noted. We refer to the beginning of the calendar year which had been observed by the Romans on the ancient date of the spring equinox, March 25 and later, March 1. In their political body they were ruled by dictators and the state was administered by various officials known as consuls. Two consuls assumed their civic duties every January 1, and because of this annual event, the date for the beginning of the civil calendar, gained in public favor. January 1 gradually superseded the March dates as new year's day and received official sanction about 153 B.C.[9] We Americans can parallel this custom with January 3 when our Congressmen take their seats and with January 20 when our Presidents are inaugurated. This change by the Romans was rather a retrogressive step in so far as the calendar was now separated from the natural seasonal year. Today we are still handicapped by having divorced the calendar from a seasonal beginning.

When Julius Caesar conquered Egypt he learned of their sun calendar which was vastly superior to the intricate moon-sun calendar used in his country. For the Roman calendar, as we have seen, was a most haphazard instrument wherein nothing could be planned with ease and accuracy. Caesar zealously applied himself to remove from the calendar all past political manipulation and to bring into it some semblance of order, accuracy and stability. That he had been elected Pontifex Maximus in 63 B.C. made his task notably easier. He had been vested with the necessary authority.

In this stupendous and obviously difficult task he wisely called to his aid the Grecian born astronomer, Sosigenes, whose work for many years had been in Alexandria. He was

---

[9] *The Encyclopaedia Britannica,* 14th Edition, Volume 4, p. 579.

thus well versed in the usage of the Egyptian calendar and apparently knew of the Canopus decree.

In the Julian revision the year was given 365¼ days and the familiar 12 months were retained. Referring to the two Latin writers, Censorinus, in his book, *The Natal Day,* written in 238 A.D. (wherein he dealt in the second part with the chronology and principal epochs of ancient history), and Macrobius c. 325-385 A.D., in his *Saturnalia* (writing on the history and discussions of the Roman calendar), we learn that the 10 new days, necessary to conform with the sun calendar, were distributed throughout the months. The quarter-day was withheld until it had reached one full day. This day was then inserted—intercalated—into the calendar every four years at the end of February, giving the quadrennial years 366 days. It was the leap-year day that Egypt several centuries before had endeavored to initiate but which the people and the priesthood had rejected. Julius Caesar had simplified the calendar.

Because, I believe, these ancient books of the third and fourth centuries A.D. are of general interest I shall quote the passages below which deal with the ten-day intercalation. My information is based on the article, "The Roman Calendar and Its Reformation by Julius Caesar," by Dr. Roscoe Lamont, published in *Popular Astronomy,* November, 1919.

From Censorinus we learn:

"The pontiffs were charged with making the intercalation, but most of them, on account of enmity or friendship, shortened or lengthened the term of a magistrate, and intercalating more or less according to their pleasure, caused a farmer of the revenue to gain or lose according to the length of the year, thus making worse what was given to them to correct. The disorder was so great that Caius Caesar, chief pontiff, in his third consulate and that of M. Aemilius Lepidus, corrected former abuses by placing two intercalary months of 67 days between November and December, although he had already intercalated 23 days in the month of February, thus giving to that year 445 days. At the same time, to prevent a like error from occurring in the future, he abandoned the intercalary month and regulated the civil year by the course of the sun, adding 10 days to the former 355, so distributed among the seven months having 29 days that two days were added to January, Sextilis and December and one to each of the others, and he placed these days at the ends of the months in order that the religious festivals might remain in their former places.... As to the quarter of a day which it seems completes the true year, he directed that one day be intercalated, after each period of four years, where the month was formerly placed, that is, after Terminalia (February 23rd), which is now called bissextile day. From this year, thus regulated by Julius Caesar, those down to our time are called Julian, and they began at the fourth consulate of Caesar."

## The version of Macrobius is:

"Julius Caesar added ten days to the former number in order to complete the 365 days which the sun takes to pass through the zodiac; and to take account of the quarter of a day, he directed the pontiffs, who were entrusted with the months and days, to intercalate one day every four years in the same month and in the same place the ancients had intercalated, that is, before the last five days of February, hence this day was called bissextile. The ten days added were distributed in such a way that January, Sextilis and December received two days, April, June, September and November one only. He added nothing to February in order that the religious rites in honor of the gods of the nether world might not be changed. March, May, Quintilis and

October remained as they were, their 31 days being a sufficient number, though they had their nones on the seventh as Numa had ordered, Caesar not having changed this arrangement. January, Sextilis and December, to which Caesar added two days, although after Caesar they began to have 31 days, still had their nones on the fifth, nineteen days being counted from the ides to the calends of the following month."

In other words, the seven months which had 29 days were changed as follows: January, August (Sextilis), December were given 31 days each, and April, June, September, November, 30 days each. The four months of 31 days—March, May, July (Quintilis), October—remained the same, and February of 28 days was unchanged. From this we see that the lengths of the calendar months were the same as they are today.

Julius Caesar was most anxious to begin the calendar on one of the seasonal positions it had enjoyed in earlier days. His choice was the winter solstice, December 25. But again the people resisted change. The traditional moon was still of greater importance to them than the new and strange sun year. By a curious coincidence, a new moon in the year of the Julian reform, 45 B.C., came on January 1, the same day the consuls took public office. To placate the populace, Caesar compromised. January 1 remained the New Year's Day, and we today are still laboring under this compromise and strange incongruity.

Caesar's desire to begin the year with the winter solstice, December 25, was a most reasonable one as it would have placed the beginning of the calendar year with the beginning of a season. It also would have had a direct association with the great feast day of pre-Christian Romans, Saturnalia,

December 25, which annually celebrated the winter solstice, that marked the sun's upward swing in the heavens
after the lowest point of the sun's downward swing had been
reached. The calendar thus would have recognized and been
in harmony with one of nature's seasonal periods.

The Julian reform adopted in 45 B.C., followed the long
year of adjustment (well called the year of confusion with
its 445 days in 46 B.C.) and marked as signal an event in
Italian history as did the sun calendar in the annals of Egypt.
Through Caesar, the sun calendar had become firmly en-
trenched. The haphazard intercalation of an entire month
had now been reduced to one day only, and the calendar
had become orderly, dependable and stable. And yet it was
not the best. The irregular lengths of months still remained.
And the complicated Roman system of counting days back-
ward within the months in the strange groupings of Calends,
Nones and Ides called for improvement. Who does not recall "Remember March, the ides of March remember," that
tragic day of a great calendar reformer?

Julius Caesar by his courage and wisdom, and willingness
to call in a scientist to assist him, gave his people a calendar
worthy of his position and name. The Roman Senate grate-
fully recognizing the remarkable achievement honored
Caesar by renaming his birth-month, Quintilis, July. This
honor, however, was bestowed upon him after his assassina-
tion and was a posthumous tribute.

With this Julian calendar in use over the vast Roman
Empire, Europe was joined to Africa in the measurement of
Time. The calendar had become a uniting link but unfor-
tunately it was not to remain so. As the calendar united so
could it also divide.

# 6

# THE AUGUSTAN AND CON-
# STANTINIAN CHANGES

THE AUGUSTAN CHANGE occurred within the Julian calendar itself without necessitating a change in the Julian reform. After the Julian revision had been in use for about a quarter-century it was discovered that once again the calendar was running away from the seasons, caused through misunderstanding by the pontiffs of the Julian leap-year rule. Too many February 29ths had been inserted so that by 8 B.C. thirteen leap years instead of the prescribed ten (according to the rule) had been given the calendar. To correct the error Augustus decreed that from 8 B.C. to 8 A.D. all leap years were to be suspended.[10] It was in 8 A.D. that the leap year reappeared in the calendar and the rule became simple and clear. Every fourth year was to be a leap year of 366 days.

It is interesting to note that the temporary withholding of leap years occurred within the transitional period of pagan chronology known as B.C. and that of the Christian, A.D. The division of the two eras into B.C. and A.D. was the work of Dionysius Exiguus, also known as Denys the Little of the sixth century. He established Christ's birth as 1 A.D. Accord-

[10] Fotheringham, Dr. John K., *Nautical Almanac,* p. 760. London, England, 1935.

ing to scientists and historians today, the birth is placed about 4 B.C. It is well to remember that Dionysius Exiguus did not use a zero year, 1 B.C. was immediately followed by 1 A.D., so that there was no intervening year between the two eras. They followed each other without a break.

Again the Roman Senate took cognizance of a calendar adjustment. It decreed the changing of the month Sextilis to that of August in the following explanatory words by Macrobius:

"Whereas the Emperor Augustus Caesar in the month of Sextilis was first admitted to the consulate and thrice entered the city in triumph, and in the same month the legions from the Janiculum placed themselves under his auspices, and in the same month Egypt was brought under the authority of the Roman people, and in the same month an end was brought to the civil wars, and whereas for these reasons the said month is and has been most fortunate to this empire, it is hereby decreed by the Senate that the said month shall be called Augustus." [11]

This senatorial account, together with the historical words of Censorinus and Macrobius previously quoted, should wipe out the questionable legend that Augustus changed the length of his month August from 30 to 31 days for egotistical motives. It has been believed that he did this so that the month August would have the same number of days as the preceding month of July which had been named in honor of his illustrious predecessor and uncle, Julius Caesar.

The legend seems to have originated in England, first in 1232 A.D. by Sacro Bosco in his book, *De anni ratione,* followed in 1812 by John Brady in his *Clavis Calendaria* or a

[11] Lamont, Dr. Roscoe, *Popular Astronomy,* November 1919; *Journal of Calendar Reform,* Volume IX, p. 11.

*Compendious Analysis of the Calendar,* and further perpetuated by Mr. Galloway in his article in the *Britannica* about 1830.[12] It was this latter article published in the scholarly *Britannica,* which enjoys a world-wide authority, that it received wide-spread acceptance.

It was only with later research that the more ancient books, *The Natal Day,* by Censorinus, and the *Saturnalia,* by Macrobius, came to light and ancient historical facts became known.

There is another group, however, which holds that Augustus did manipulate the calendar, whereby the month of August would have the coveted 31 days the same as July, and that he took a day from February and readjusted September and November to have 30 days each, and October and December 31 days each.

In the recording of ancient history it is inevitable that much remains clouded and hidden. Different interpretations and deductions by various minds and nationalities, too, are unavoidable and it is open to question whether these differences are reconcilable. In most instances such divergencies do not interfere with the present or the future, as the past is irrevocably past and irretrievable. In dealing with the ancient past it would seem that a spirit of tolerance and fairmindedness should prevail, so that similar errors are not made and perpetuated. One cannot be too positive or too certain about the ancient past where records vary and are incomplete.

[12] Lamont, Dr. Roscoe, *Popular Astronomy,* November 1919.

# CONSTANTINIAN CALENDAR

## EACH YEAR DIFFERENT

|     | S | M | T | W | T | F | S |
|-----|---|---|---|---|---|---|---|
| **JAN** | 1 | 2 | 3 | 4 | 5 | 6 | 7 |
|     | 8 | 9 | 10 | 11 | 12 | 13 | 14 |
|     | 15 | 16 | 17 | 18 | 19 | 20 | 21 |
|     | 22 | 23 | 24 | 25 | 26 | 27 | 28 |
|     | 29 | 30 | 31 |   |   |   |   |
| **FEB** |   |   |   | 1 | 2 | 3 | 4 |
|     | 5 | 6 | 7 | 8 | 9 | 10 | 11 |
|     | 12 | 13 | 14 | 15 | 16 | 17 | 18 |
|     | 19 | 20 | 21 | 22 | 23 | 24 | 25 |
|     | 26 | 27 | 28 | (29 in leap year) |   |   |   |
| **MAR** |   |   |   | 1 | 2 | 3 | 4 |
|     | 5 | 6 | 7 | 8 | 9 | 10 | 11 |
|     | 12 | 13 | 14 | 15 | 16 | 17 | 18 |
|     | 19 | 20 | 21 | 22 | 23 | 24 | 25 |
|     | 26 | 27 | 28 | 29 | 30 | 31 |   |
| **APR** |   |   |   |   |   |   | 1 |
|     | 2 | 3 | 4 | 5 | 6 | 7 | 8 |
|     | 9 | 10 | 11 | 12 | 13 | 14 | 15 |
|     | 16 | 17 | 18 | 19 | 20 | 21 | 22 |
|     | 23 | 24 | 25 | 26 | 27 | 28 | 29 |
|     | 30 |   |   |   |   |   |   |
| **MAY** |   | 1 | 2 | 3 | 4 | 5 | 6 |
|     | 7 | 8 | 9 | 10 | 11 | 12 | 13 |
|     | 14 | 15 | 16 | 17 | 18 | 19 | 20 |
|     | 21 | 22 | 23 | 24 | 25 | 26 | 27 |
|     | 28 | 29 | 30 | 31 |   |   |   |
| **JUN** |   |   |   |   | 1 | 2 | 3 |
|     | 4 | 5 | 6 | 7 | 8 | 9 | 10 |
|     | 11 | 12 | 13 | 14 | 15 | 16 | 17 |
|     | 18 | 19 | 20 | 21 | 22 | 23 | 24 |
|     | 25 | 26 | 27 | 28 | 29 | 30 |   |

|     | S | M | T | W | T | F | S |
|-----|---|---|---|---|---|---|---|
| **JUL** |   |   |   |   |   |   | 1 |
|     | 2 | 3 | 4 | 5 | 6 | 7 | 8 |
|     | 9 | 10 | 11 | 12 | 13 | 14 | 15 |
|     | 16 | 17 | 18 | 19 | 20 | 21 | 22 |
|     | 23 | 24 | 25 | 26 | 27 | 28 | 29 |
|     | 30 | 31 |   |   |   |   |   |
| **AUG** |   |   | 1 | 2 | 3 | 4 | 5 |
|     | 6 | 7 | 8 | 9 | 10 | 11 | 12 |
|     | 13 | 14 | 15 | 16 | 17 | 18 | 19 |
|     | 20 | 21 | 22 | 23 | 24 | 25 | 26 |
|     | 27 | 28 | 29 | 30 | 31 |   |   |
| **SEP** |   |   |   |   |   | 1 | 2 |
|     | 3 | 4 | 5 | 6 | 7 | 8 | 9 |
|     | 10 | 11 | 12 | 13 | 14 | 15 | 16 |
|     | 17 | 18 | 19 | 20 | 21 | 22 | 23 |
|     | 24 | 25 | 26 | 27 | 28 | 29 | 30 |
| **OCT** | 1 | 2 | 3 | 4 | 5 | 6 | 7 |
|     | 8 | 9 | 10 | 11 | 12 | 13 | 14 |
|     | 15 | 16 | 17 | 18 | 19 | 20 | 21 |
|     | 22 | 23 | 24 | 25 | 26 | 27 | 28 |
|     | 29 | 30 | 31 |   |   |   |   |
| **NOV** |   |   |   | 1 | 2 | 3 | 4 |
|     | 5 | 6 | 7 | 8 | 9 | 10 | 11 |
|     | 12 | 13 | 14 | 15 | 16 | 17 | 18 |
|     | 19 | 20 | 21 | 22 | 23 | 24 | 25 |
|     | 26 | 27 | 28 | 29 | 30 |   |   |
| **DEC** |   |   |   |   |   | 1 | 2 |
|     | 3 | 4 | 5 | 6 | 7 | 8 | 9 |
|     | 10 | 11 | 12 | 13 | 14 | 15 | 16 |
|     | 17 | 18 | 19 | 20 | 21 | 22 | 23 |
|     | 24 | 25 | 26 | 27 | 28 | 29 | 30 |
|     | 31 |   |   |   |   |   |   |

THIS CALENDAR HAS 52 WEEKS AND MUST BORROW ONE DAY FROM ANOTHER WEEK TO COMPLETE THE YEAR. THIS CAUSES THE CALENDAR TO CHANGE EVERY YEAR AND IS RESPONSIBLE FOR ITS CONFUSION.

## CONSTANTINIAN CHANGE

In 321 A.D., a far-reaching change was made within the arrangement of the calendar itself without in any way altering its scientific accuracy. The change was an internal one, its far-reaching significance not grasped at the time.

Emperor Constantine the Great had been converted to the new religion that was sweeping Europe. He had become a Christian. Wishing to honor the Christ and to accept the seven-day week with its regular one day of rest, he officially decreed the week as a part of the calendar, at the same time that he designated the first day of every week—Sunday —the Christian day of rest, prayer and worship. The first day was chosen by the Emperor in acknowledgment of the day upon which Christ rose from the dead, revealing immortality. The week had become Christianized.

It is generally accepted that the Hebrews and Babylonians gave us the seven-day week, as we know it. To the Hebrew the week symbolized not only the six vast periods of creation but also the Sabbath, the seventh period of rest, when the Creator ceased from His labors. To the Hebrew the Sabbath, and to the Christian the Sunday, are days of religious worship, and appear to have more significance than the mooted unbroken sequence of the seven-day weeks, which lacks historical authority. Dr. Julian Morgenstern, president of The Hebrew Union College of Cincinnati writes searchingly on the history of the ancient Israel calendars, from which I quote:

"The Bible furnishes ample evidence that, at different periods in ancient Israel, three different festival calendars and calendar systems were employed. . . .

"It is generally taken for granted,...the calendars agreed in all essential respects, except in the manner by which they indicated the month....For, surely, between names of Canaanite origin and names of Babylonian origin ancient Israel could have had little preference. This consideration suggests the probability that these two changes of the manner of designating the months in ancient Israel may have been due to causes of *considerable importance,* and may have been accompanied by *internal revision of the entire calendar system of more than passing significance.* "...there were three different calendars employed at different times in ancient Israel, and that the transition from the one system of designating months to the second system, and from this in turn to a third system, implied two revisions of the calendar, *each in all likelihood, of a thorough-going nature.*" [13]

The week calls to mind the origin of the names of its seven days. The ancients called the weekdays after the then known seven planets which include the sun and the moon— Sun, Moon, Mars, Mercury, Jupiter, Venus and Saturn. It is most interesting how the various countries have adapted these names to their language or, as in the case of the Anglo-Saxon, substituted some pagan gods in place of the planets.

| LATIN | FRENCH | SPANISH | ANGLO-SAXON |
|---|---|---|---|
| Dies Solis | Dimanche | Domingo | Sunnan-daeg |
| Dies Lunae | Lundi | Lunes | Monan-daeg |
| Dies Martis | Mardi | Martes | Tiwes-daeg |
| Dies Mercurii | Mercredi | Miercoles | Wodnes-daeg |
| Dies Jovis | Jeudi | Jueves | Thunres-daeg |
| Dies Veneris | Vendredi | Viernes | Frige-daeg |
| Dies Saturni | Samedi | Sabado | Saetern-daeg |

[13] Morgenstern, Julian, *The Three Calendars of Ancient Israel,* pp. 13-78. Hebrew Union College, Annal for 1924, Cincinnati, Ohio, U.S.A. (Italics mine, E. A.)

The future holds the promise that all peoples of whatever race, creed, nation, or culture will ultimately observe in their daily lives one day of rest in the week. It is this one day of rest which provides man with the spiritual outlook as he reminds himself from whence he came, after whose likeness he was made, and the divine promise of his ultimate return to his Creator.

The week of seven days brought incalculable benefits to the social, cultural, physical and religious life of man. But to the calendar, the week proved more of a disadvantage than an advantage. Mathematically the calendar became disrupted, since seven days repeated fifty-two times do not fit into the sun calendar of 365 days with an occasional 366th day. This series of fifty-two weeks is always short one day at the conclusion of every year, so that the calendar must borrow one day from the following week. It is this shortage that causes the civil calendar to be variable every year. The calendar because of the week, lost its previous stability. The week is so beneficial in our civil, social and religious life that we would not wish to forego it; however, the changeability and irregularity that the week brought to the calendar is deplorable. We shall see shortly the means by which the lost order can be regained and the needed stability restored without in any way sacrificing the seven-day week with its one day of rest, prayer and worship.

# 7
# THE GREGORIAN CALENDAR

Twelve hundred and sixty-one years passed after the introduction of the week into the civil calendar before another change was effected. It is the one we know today as the Gregorian calendar and generally used throughout the world. During this long interval the calendar had been put to a rigid test as to its scientific accuracy and religious adaptability. The practical workability pertaining to the civil, social and cultural aspects, which so distinguished the earlier Egyptian calendar, was still of lesser concern to revisionists. Science and religion claimed their full privilege and authority.

Scientists had discovered with the passing of Time that it did not take the earth an exact number of 365¼ days to complete her annual circuit around the sun, but a fraction less than a quarter-day. The length was actually 365 days, 5 hours, 48 minutes and 46 seconds. In other words, the length was 365.2422 days. This slight fractional difference caused the calendar once again to fall out of step with the seasons.

At the time of the Gregorian reform in 1582 A.D. the calendar had run ahead by 14 days, according to Julian reckon-

# GREGORIAN CALENDAR

## EACH YEAR DIFFERENT

| | S | M | T | W | T | F | S |
|---|---|---|---|---|---|---|---|
| **JAN** | | 1 | 2 | 3 | 4 | 5 | 6 |
| | 7 | 8 | 9 | 10 | 11 | 12 | 13 |
| | 14 | 15 | 16 | 17 | 18 | 19 | 20 |
| | 21 | 22 | 23 | 24 | 25 | 26 | 27 |
| | 28 | 29 | 30 | 31 | | | |
| **FEB** | | | | | | 1 | 2 | 3 |
| | 4 | 5 | 6 | 7 | 8 | 9 | 10 |
| | 11 | 12 | 13 | 14 | 15 | 16 | 17 |
| | 18 | 19 | 20 | 21 | 22 | 23 | 24 |
| | 25 | 26 | 27 | 28 | ★ | | |
| **MAR** | | | | | 1 | 2 | 3 |
| | 4 | 5 | 6 | 7 | 8 | 9 | 10 |
| | 11 | 12 | 13 | 14 | 15 | 16 | 17 |
| | 18 | 19 | 20 | (21) | 22 | 23 | 24 |
| | 25 | 26 | 27 | 28 | 29 | 30 | 31 |
| **APR** | 1 | 2 | 3 | 4 | 5 | 6 | 7 |
| | 8 | 9 | 10 | 11 | 12 | 13 | 14 |
| | 15 | 16 | 17 | 18 | 19 | 20 | 21 |
| | 22 | 23 | 24 | 25 | 26 | 27 | 28 |
| | 29 | 30 | | | | | |
| **MAY** | | 1 | 2 | 3 | 4 | 5 |
| | 6 | 7 | 8 | 9 | 10 | 11 | 12 |
| | 13 | 14 | 15 | 16 | 17 | 18 | 19 |
| | 20 | 21 | 22 | 23 | 24 | 25 | 26 |
| | 27 | 28 | 29 | 30 | 31 | | |
| **JUN** | | | | | | 1 | 2 |
| | 3 | 4 | 5 | 6 | 7 | 8 | 9 |
| | 10 | 11 | 12 | 13 | 14 | 15 | 16 |
| | 17 | 18 | 19 | 20 | 21 | 22 | 23 |
| | 24 | 25 | 26 | 27 | 28 | 29 | 30 |

| | S | M | T | W | T | F | S |
|---|---|---|---|---|---|---|---|
| **JUL** | 1 | 2 | 3 | 4 | 5 | 6 | 7 |
| | 8 | 9 | 10 | 11 | 12 | 13 | 14 |
| | 15 | 16 | 17 | 18 | 19 | 20 | 21 |
| | 22 | 23 | 24 | 25 | 26 | 27 | 28 |
| | 29 | 30 | 31 | | | | |
| **AUG** | | | 1 | 2 | 3 | 4 |
| | 5 | 6 | 7 | 8 | 9 | 10 | 11 |
| | 12 | 13 | 14 | 15 | 16 | 17 | 18 |
| | 19 | 20 | 21 | 22 | 23 | 24 | 25 |
| | 26 | 27 | 28 | 29 | 30 | 31 | |
| **SEP** | | | | | | | 1 |
| | 2 | 3 | 4 | 5 | 6 | 7 | 8 |
| | 9 | 10 | 11 | 12 | 13 | 14 | 15 |
| | 16 | 17 | 18 | 19 | 20 | 21 | 22 |
| | 23 | 24 | 25 | 26 | 27 | 28 | 29 |
| | 30 | | | | | | |
| **OCT** | | 1 | 2 | 3 | 4 | 5 | 6 |
| | 7 | 8 | 9 | 10 | 11 | 12 | 13 |
| | 14 | 15 | 16 | 17 | 18 | 19 | 20 |
| | 21 | 22 | 23 | 24 | 25 | 26 | 27 |
| | 28 | 29 | 30 | 31 | | | |
| **NOV** | | | | | | 1 | 2 | 3 |
| | 4 | 5 | 6 | 7 | 8 | 9 | 10 |
| | 11 | 12 | 13 | 14 | 15 | 16 | 17 |
| | 18 | 19 | 20 | 21 | 22 | 23 | 24 |
| | 25 | 26 | 27 | 28 | 29 | 30 | |
| **DEC** | | | | | | | 1 |
| | 2 | 3 | 4 | 5 | 6 | 7 | 8 |
| | 9 | 10 | 11 | 12 | 13 | 14 | 15 |
| | 16 | 17 | 18 | 19 | 20 | 21 | 22 |
| | 23 | 24 | 25 | 26 | 27 | 28 | 29 |
| | 30 | 31 | | | | | |

THIS CALENDAR HAS 52 WEEKS AND MUST BORROW ONE DAY FROM ANOTHER WEEK TO COMPLETE THE YEAR. THIS CAUSES THE CALENDAR TO CHANGE EVERY YEAR AND IS RESPONSIBLE FOR ITS CONFUSION.

ing; the spring equinox instead of coming on March 25 appeared on March 11. Were this condition to continue uncorrected, we should find ourselves ultimately in a most ludicrous position. We should be observing spring in the winter months and winter in the autumn months; and likewise autumn in the summer months and summer in the spring months. Eventually the seasons would pass through all of the twelve months in the Julian calendar, as the Mohammedan moon calendar does today; in 34 years the seasons complete their full round. As this moon calendar, however, is primarily used by peoples living in the tropical zone where the seasons are but two—wet and dry—this strange situation is possible.

With the seasonal complications experienced in the Julian calendar, the religious feast days, too, were suffering since their significance was ofttimes lost because of the incongruous seasons in which they appeared. Just imagine were Christmas to be observed in autumn or spring! Something quite drastic had to be done.

Pope Gregory XIII wisely took charge of the situation. For now it was not a pre-Christian Roman Pontiff but a Christian Pope who was living in Rome and who had automatically succeeded as Pontifex Maximus, thereby assuming control of the calendar. Christianity having split into various factions, the Eastern Orthodox Church with its seat in Constantinople and the Protestant groups with their various organizations throughout Europe, caused the Roman Catholic Church in the sixteenth century to be the most powerful and influential of Christian Churches.

The Pope in effecting his reform had two advisers in the conference which took place in the Gregorian Tower at the

Vatican in 1582: Aloysius Lilius, a learned astronomer and physician of Naples, who devised the new system but did not survive its adoption, and Christopher Clavius, a German Jesuit and mathematician.

The Gregorian reform consisted of two amendments and two reaffirmations.

The two amendments were as follows:

The Julian calendar in arrangement was not changed, but the date of the spring equinox was adjusted. Instead of reverting to the pre-Christian date of the spring equinox, March 25, Pope Gregory XIII based the equinox on the date when it had occurred in 325 A.D., the year of the first great Christian assembly that had gathered at Nicea. The equinox in that year had occurred on March 21. In 1582, when the Gregorian reform was adopted, March 21 was selected as the spring equinox for the entire Christian world to commemorate the Council of Nicea. The calendar had now become "Christianized" as the week had previously become "Christianized" under Constantine the Great.

To effect this change, *ten days* had to be dropped from the calendar—days coming between October 4 and 15 in 1582—without disrupting the sequence of the weekdays: Thursday, October 4, was followed by Friday, October 15. Imagine going to bed Thursday evening, October 4, and on awakening the next morning, Friday, October 15, find that the calendar had made you ten days older!

In order to retain March 21 for future spring equinoxes, the former Julian leap-year rule required amendment. The new leap-year rule stated that century years divisible in whole numbers by 400 were to be leap years. Thus 1600 was a leap year but not 1700, 1800 and 1900, whereas 2000 will again be a leap year. All other years retained the simple four-year rule.

The two reaffirmations were those of New Year's Day and Easter. New Year's Day was retained as of January first of ancient Roman origin. Easter, as established by the Council of Nicea, was reaffirmed as the first Sunday after the full moon following the spring equinox. The Gregorian reform still retained a curious contradiction; the civil calendar was reckoned by the sun, and Easter, a great religious feast day, by the moon.

At the time of the reform, we are told, certain Catholic priests pointed out how extremely desirable it would be to fix the date for Easter definitely while reforming the calendar. And the great leader of the Reformation, Martin Luther, also did not agree with the regulations then in use. He wrote:

"How much better it would have been to have dropped entirely the Mosaic law referring to Easter, rather than patch it up. They should have laid down a *special* date for the Passion, death and Resurrection, just as has been done for Christmas, Epiphany, Candlemas and other feasts which are fixed and not movable, so that everybody might know exactly, and without trouble or dispute, the dates of Easter and the feasts depending on Easter. As things are now, we really need a further reform to correct the calendar and fix Easter." [14]

Notwithstanding these justifiable criticisms the sixteenth century reform was notable. It adjusted the calendar to the seasons so that with this "new style" 3,323 years (from its adoption, 1582 A.D.) will have to pass before the calendar

[14] *The Churches and the Stabilisation of Easter,* p. 26. International Christian Social Institute, Geneva, 1933.

accumulates one day too many, so that in 4905 A.D. a correction of one day will have to be made. No one need worry, however, for when the proper time arrives it will be easy to correct, as scientists have pointed out, by the simple method of considering a leap year as an ordinary year.

The Gregorian reform, excellent in so many ways, failed to receive approval among nations. Opposition arose because one Christian group dealt with the matter single-handed and alone, without consulting and co-operating with other Christian groups. This lack of consideration was lamentable. Countries, among them Protestant England and the Eastern Orthodox nations, refused to accept any "Papal" calendar, fearful of losing their individuality and becoming subject to Rome. Thus only Roman Catholic countries accepted it in 1582. Protestant England adopted it as late as 1752, Japan in 1873, China in 1912, Soviet Russia in 1918 (changed October 1, 1929, to one of her own and then on June 26, 1940, returned to the regular seven-day week),[15] Eastern Orthodox Greece 1924, and Turkey in 1927. There are still Eastern Orthodox groups today who tenaciously cling to the older Julian calendar which lags 13 days behind that of the Gregorian. Their Christmas, December 25, occurs on January 7 in the "new style" Gregorian.

To the United States of America, the approval and adoption of the Gregorian calendar by Great Britain in 1752, 170 years later, brought about a curious complication of a national holiday. Under the Julian reckoning, "old style," George Washington was born February 11, 1731, but with the "new style" his birthday became February 22, 1732. Washington had the unique distinction officially to observe two birthdays

[15] *Bulletin,* July 5, 1940, American Russian Institute, Inc.

—the original and the revised. The year of his birth also was altered in the new style reckoning, as designated in the Washington family Bible at Mt. Vernon.[16] This interesting entry reads:

"George Washington, son to Augustine and Mary his wife, was born ye 11th day of February 1731/2 about 10 in the morning, and was baptized on the 30th of April following."

Because of the unfortunate opposition that had arisen, it was really obligatory for the Gregorian calendar to have several centuries in which to entrench itself firmly in the pattern of our everyday life. Too many changes in the calendar following the adoption would have undoubtedly caused confusion, weakening the foundation upon which it was built. It was not until the twentieth century—our century—that the Gregorian calendar became the generally accepted time-system of the world. Only then were the serious defects more widely felt and careful thought directed to the calendar's further improvement. Several attempts had been made in the nineteenth century but these appear to have been at best trial balloons, exploring, testing and obtaining valuable information, all necessary preparation for a change to be effected later.

[16] Although the official Julian New Year's was January 1, certain countries like Great Britain began their calendar year with the ancient spring equinox date, March 25. February of that year was in British reckoning, 1731; in the Gregorian calendar 1732. Owing to a centurial leap year, 1600, having intervened since the Gregorian reform, Great Britain had to drop 11 not 10 days from the calendar.

# 8

# PROGRESS REVEALS DEFECTS

Significant events appear in every century when outstanding discoveries, notable inventions, and inspiring ideas spur man onward in his unending quest toward perfection. Each century stands forth with its own particular contribution. Retracing our steps for only five centuries we observe that the fifteenth was notable for Gutenberg's invention of the printing press making knowledge available to the people heretofore reserved for special groups; the sixteenth gave us the Gregorian calendar still in use today; the seventeenth witnessed the colonization of the Western Hemisphere whereby it became more closely linked to the Eastern. The eighteenth century inaugurated a new form of government—a democracy founded on two indestructible pillars, man's freedom and his inalienable rights; the nineteenth brought the outstanding discovery of electricity; and the twentieth—our century—ushered in aviation and the radio. Our planet is rapidly being transformed into a closely related globe on which we live, work, play and worship. It can no longer be considered as separated in isolated and disorganized groups.

Recognizing this, it is natural that Earth's rhythmic

breathing should also guide and uphold man in his calendar system. To disregard it might seriously hinder his reasonable ambition to reach out toward better things, to establish better conditions everywhere. For is not everything predicated on Time? Time which is known as a healer, can truly heal when the calendar is healed.

The printing press of the fifteenth century was as important a step in the emancipation of man from the slavery of ignorance into the open free field of knowledge as was the Flavian act that preceded the Christian era. It is significant that among the lesser known books printed by Gutenberg and his associates, was the *Astronomical Kalendar* set in the famous 36-line Bible type, ascribed to the year 1447, of which today but a fragment on vellum remains. Here again we are impressed with the effect of Time upon man's consciousness. Ever is he aware of Time's mysterious influence.

The invention of the printing press allowed people to become familiar with the calendar. It emerged from its previous seclusion and became the property of all, so that today any contemplated change of the civil calendar is momentous to everyone; to governmental, business, social, cultural and educational groups, as well as to science and religion.

Before entering upon the question, "What type of calendar should the new one be?" let us outline the defects of the present calendar. Let us clearly understand that the calendar is far from perfect but the length of the year is as accurate as we could wish. The valuable work accomplished by scientists and scholars in the past stand secure. The sun calendar of 365 days with an occasional 366th leap-year day and divided into the familiar 12 months remains an enduring guidepost along man's long march of development from the dim past into the better-known present, out into the unknown, hopeful future.

The present calendar's greatest defect is undeniably its unceasing changeability. No two like years follow each other. In fact there are fourteen different types of calendar years (seven average and seven leap years—one year for each day of the week) so that no one can tell from year to year which of the fourteen types is in use in any particular year. We have four types of months: 28, 29, 30, 31 days. Each of these months can begin on any of the seven days of the week, so that there are, all in all, 28 different months. Our calendar is a system of "confusion worse confounded."

For instance, we never know how many Sundays, Mondays, Tuesdays, Wednesdays, Thursdays, Fridays or Saturdays, January had last year, this year, or will have next year. The same is true of the other months. The quarter and half-years also contribute their share to the uncertainty as the year is unevenly divided into quarters of 90, 91, 92 days, and the half-years into 181, 182, 184 days. Who can tell the lengths of our 12 months without silently, ofttimes shame-facedly, repeating that silly yet indispensable little nursery rhyme?

> "Thirty days hath September,
> April, June and November;
> All the rest have thirty-one,
> Excepting February alone,
> Which hath but twenty-eight, in fine,
> Till leap year gives it twenty-nine."

Blessed is the unknown author who penned these lines for our daily guide, whether for children, youths, adults or the aged. In many countries the knuckles of one hand are used to designate the 31-day months.

Because of the planlessness of our time-system, with month-dates and weekdays never agreeing from year to year (Christmas, December 25, can fall on any of the seven days of the week), is it surprising that our calendar contributes to the disorder and confusion in our lives? Would one go off on a journey by train, boat, automobile, plane or hiking without knowing some of the stops and routes to guide one's destination? No one in reason would undertake any worthwhile trip without knowing something about the plan.

Would anyone use a watch that lags, or skips, or runs ahead, in other words is unreliable? It would not be kept a single day. One would either see to it that it regains its lost regularity or it would be discarded for a better one. Well, that is what this century and decade are going to do with the present calendar. They are no longer content to endure its planless, irregular and inconvenient arrangement. Just because we are familiar with the calendar is no reason for clinging to it blindly.

Here is an example that applies to the year 1943 in the United States. In our present calendar of unequal quarters let us compare the first quarter of 90 days with the second quarter of 91 days. We discover that January has 8½ non-productive or overtime wage days consisting of 5 half-day Saturdays, five Sundays and, at least, one holiday. February has eight such overtime days and March has six for which overtime wages are paid. There are thus 22½ overtime wage days out of the 90 days. In like manner the second quarter of 91 days has 20½ non-productive or overtime wage days, so that the difference between these two quarters is two days.

Unless this difference is given careful consideration the result is either profit or loss. This also applies to the third and fourth quarters with their different holidays, Saturdays and Sundays. And of course, every calendar year is different.

And again we cannot imagine changing our currency to different dollars and cents every year, buying clocks with different hours, tape measures with different arrangements of inches, and scales with different sets of pounds and ounces.

What profligates we are regarding the paper, printer's ink, time, effort and money that we expend every year getting out a new and different calendar! No one should be so foolish as to believe that the calendar cannot be changed. The urgencies of war make increased efficiency necessary, and are changing many things, why not the calendar? It surely would be an incalculable saving and benefit were the calendar to be improved to fit present emergencies and future requirements; no longer sabotaging valuable time, effort, money and material.

War days and ensuing peace days demand that we use systems and schedules that are well planned, dependable, and in which we can—each one of us—do our best work in the best possible time with the least possible waste.

The military has given us a new and valuable word— logistics. Thomas J. Watson, president of the International Business Machines Corporation, defines logistics admirably:

"to do the best possible job at the lowest possible price and in the shortest possible time." [17]

If the military functions logistically and business is being

[17] *Proceedings of the Academy of Political Science,* Volume XX, May 1942, Number 1, p. 60, by permission of Mr. Watson.

geared logistically, why not a calendar planned logistically?

Our present calendar is bad logistics. It has no well-laid plan, no co-ordination. One year begins on a Sunday, another on a Monday, a Wednesday, a Thursday and so on. To have months with 28 or 31 days is also unfair. Midweek holidays are a nuisance, interfering with the smooth flow of everyday life, particularly with business and education. They fail utterly to give one that honest-to-goodness-holiday feeling which always comes when a holiday falls on a week-end. Our calendar instead of being the best system for our daily use is actually a detriment. And *now,* when every second and minute, every hour and day are so valuable, that we should still persist in using such a senseless time-system is not only unjustified, but is a wrong that demands correction. Mathematically, our calendar today is a crazy patchwork wherein nothing fits. Our Earth is sick with its labored, hectic, irregular breathing.

The world needs desperately a new time-plan that will bring economy out of waste, order out of disorder, regularity out of confusion, and harmony out of discord. It should be a practical and serviceable instrument for daily use to accomplish the many tasks efficiently and well.

> "The flow of life cannot be stopped,
>   Neither as a whole, nor in its parts.
> It shall continue forever, throwing up new forms,
>   new institutions, new experiences.
> And every form and phase shall be discarded, as soon
>   as it becomes outworn, and of no further use;
> Both in life as a whole, and in individuals." [18]

[18] Townshend, Frank, *Earth,* p. 99, Alfred A. Knopf, Inc., N. Y., and George Allen & Unwin Ltd., London.

# THIRTEEN-MONTH CALENDAR

### JANUARY
| S | M | T | W | T | F | S |
|---|---|---|---|---|---|---|
| 1 | 2 | 3 | 4 | 5 | 6 | 7 |
| 8 | 9 | 10 | 11 | 12 | 13 | 14 |
| 15 | 16 | 17 | 18 | 19 | 20 | 21 |
| 22 | 23 | 24 | 25 | 26 | 27 | 28 |

### FEBRUARY
| S | M | T | W | T | F | S |
|---|---|---|---|---|---|---|
| 1 | 2 | 3 | 4 | 5 | 6 | 7 |
| 8 | 9 | 10 | 11 | 12 | 13 | 14 |
| 15 | 16 | 17 | 18 | 19 | 20 | 21 |
| 22 | 23 | 24 | 25 | 26 | 27 | 28 |

### MARCH
| S | M | T | W | T | F | S |
|---|---|---|---|---|---|---|
| 1 | 2 | 3 | 4 | 5 | 6 | 7 |
| 8 | 9 | 10 | 11 | 12 | 13 | 14 |
| 15 | 16 | 17 | 18 | 19 | 20 | 21 |
| 22 | 23 | 24 | 25 | 26 | 27 | 28 |

### APRIL
| S | M | T | W | T | F | S |
|---|---|---|---|---|---|---|
| 1 | 2 | 3 | 4 | 5 | 6 | 7 |
| 8 | 9 | 10 | 11 | 12 | 13 | 14 |
| 15 | 16 | 17 | 18 | 19 | 20 | 21 |
| 22 | 23 | 24 | 25 | 26 | 27 | 28 |

### MAY
| S | M | T | W | T | F | S |
|---|---|---|---|---|---|---|
| 1 | 2 | 3 | 4 | 5 | 6 | 7 |
| 8 | 9 | 10 | 11 | 12 | 13 | 14 |
| 15 | 16 | 17 | 18 | 19 | 20 | 21 |
| 22 | 23 | 24 | 25 | 26 | 27 | 28 |

### JUNE
| S | M | T | W | T | F | S |
|---|---|---|---|---|---|---|
| 1 | 2 | 3 | 4 | 5 | 6 | 7 |
| 8 | 9 | 10 | 11 | 12 | 13 | 14 |
| 15 | 16 | 17 | 18 | 19 | 20 | 21 |
| 22 | 23 | 24 | 25 | 26 | 27 | 28 |

## LEAP DAY
### June 29

### SOL
| S | M | T | W | T | F | S |
|---|---|---|---|---|---|---|
| 1 | 2 | 3 | 4 | 5 | 6 | 7 |
| 8 | 9 | 10 | 11 | 12 | 13 | 14 |
| 15 | 16 | 17 | 18 | 19 | 20 | 21 |
| 22 | 23 | 24 | 25 | 26 | 27 | 28 |

### JULY
| S | M | T | W | T | F | S |
|---|---|---|---|---|---|---|
| 1 | 2 | 3 | 4 | 5 | 6 | 7 |
| 8 | 9 | 10 | 11 | 12 | 13 | 14 |
| 15 | 16 | 17 | 18 | 19 | 20 | 21 |
| 22 | 23 | 24 | 25 | 26 | 27 | 28 |

### AUGUST
| S | M | T | W | T | F | S |
|---|---|---|---|---|---|---|
| 1 | 2 | 3 | 4 | 5 | 6 | 7 |
| 8 | 9 | 10 | 11 | 12 | 13 | 14 |
| 15 | 16 | 17 | 18 | 19 | 20 | 21 |
| 22 | 23 | 24 | 25 | 26 | 27 | 28 |

### SEPTEMBER
| S | M | T | W | T | F | S |
|---|---|---|---|---|---|---|
| 1 | 2 | 3 | 4 | 5 | 6 | 7 |
| 8 | 9 | 10 | 11 | 12 | 13 | 14 |
| 15 | 16 | 17 | 18 | 19 | 20 | 21 |
| 22 | 23 | 24 | 25 | 26 | 27 | 28 |

### OCTOBER
| S | M | T | W | T | F | S |
|---|---|---|---|---|---|---|
| 1 | 2 | 3 | 4 | 5 | 6 | 7 |
| 8 | 9 | 10 | 11 | 12 | 13 | 14 |
| 15 | 16 | 17 | 18 | 19 | 20 | 21 |
| 22 | 23 | 24 | 25 | 26 | 27 | 28 |

### NOVEMBER
| S | M | T | W | T | F | S |
|---|---|---|---|---|---|---|
| 1 | 2 | 3 | 4 | 5 | 6 | 7 |
| 8 | 9 | 10 | 11 | 12 | 13 | 14 |
| 15 | 16 | 17 | 18 | 19 | 20 | 21 |
| 22 | 23 | 24 | 25 | 26 | 27 | 28 |

### DECEMBER
| S | M | T | W | T | F | S |
|---|---|---|---|---|---|---|
| 1 | 2 | 3 | 4 | 5 | 6 | 7 |
| 8 | 9 | 10 | 11 | 12 | 13 | 14 |
| 15 | 16 | 17 | 18 | 19 | 20 | 21 |
| 22 | 23 | 24 | 25 | 26 | 27 | 28 |

## YEAR DAY
### DECEMBER 29

# 9

# THE THIRTEEN-MONTH
# CALENDAR

In the nineteenth century various attempts had been made to improve the Gregorian calendar. Among them was the proposal of Abbé Mastrofini, who in 1834 A.D. originated the idea of a 364-day year with the 365th day intercalated at the end of every year as a blank day, and in leap years the 366th day was similarly treated as another blank day. These "blank" days belonged within the calendar year yet were outside the seven-day week. This ingenious plan expressed in a booklet by the Italian Roman Catholic priest had received three *Nihil Obstats* and two *Imprimaturs* from his bishop. It was the hidden key which gave to the calendar its stability; every year would be the same in arrangement.

The 13-month calendar was devised by Auguste Comte in 1849. It, too, was a 364-day year and included the one or two blank or stabilizing days, and to each of the 13 months were given 28 days of exactly four weeks. The 13th month was inserted between June and July. This Positivist calendar was belabored with a bewilderment of names of prominent men from all times after whom the days, weeks and months were designated. The blank or 365th day every year called "Extra Day" was devoted to "all the dead" and

53

the additional 366th day in leap years called "Additional Day" was dedicated to "eminent women." It denied calendar simplification. Many years later the plan was revived by Paul Delaporte and also by Moses B. Cotsworth, both eliminating the superfluous names, which considerably simplified the plan.

A contest offered by the French Astronomical Society under the presidency of Camille Flammarion in 1887 gave an award to the best calendar plan submitted, the winner being M. Armelin. His proposal was a 12-month revision of equal quarters, having 31, 30, 30 days. Each quarter began on *Monday* and ended on *Sunday*—an arrangement, however, which gave to the first month 27 weekdays, to the second month 26 weekdays and to the third month 25 weekdays, plus Sundays. These sincere attempts toward improving the unsatisfactory calendar paved the way for more intensive and concerted action in the twentieth century.

Moses B. Cotsworth, an English Yorkshireman, born December 1859, and a Canadian (British Columbia) in later years, was undeniably the outstanding crusader for the 13-month calendar. During more than forty years he advocated this type of reform and was the first to have effectively aroused the people to the glaring defects of the present calendar. The many drastic changes, by which he proposed to remedy these, drew attention to the reform everywhere. It was so entirely different that people were arrested by its novelty. It appeared simple. The swing from the outworn old to an untried new had a most startling effect; it aroused the people to do something about improving the calendar, which in arrangement is two thousand years old.

The efforts of Mr. Cotsworth were directed mainly to the

economic and statistical fields in which he was well versed, as he was a professional accountant, knowing by experience the difficulties and inconveniences that beset all those of his profession. With unfailing confidence and zeal he visited many countries, coming to the United States in 1923. At that time the American Section of the International Chamber of Commerce had a committee to study calendar reform with which Mr. Cotsworth worked very closely. The American Standards Association had also formed a committee on the subject, of which A. Cressy Morrison was chairman. The first committee approved calendar reform but did not commit itself to any particular plan, whereas the latter, after giving the subject careful consideration, concluded it would not participate further in the matter.

Mr. Morrison, having become very much interested in Mr. Cotsworth's work, arranged a meeting between the late George Eastman, kodak manufacturer and philanthropist, and Mr. Cotsworth. This proved of inestimable value. Mr. Eastman became his staunch patron and with his moral and financial support the International Fixed Calendar League in London was formed with Mr. Cotsworth as the European director. Another and an *unofficial* National Committee on Calendar Simplification in the United States was established at Rochester, N. Y., of which George Eastman was chairman and the late Meredith N. Stiles, director. Mr. Cotsworth had now won the much-prized authoritative endorsement by an outstanding and influential American. Modern calendar revision had been successfully launched, and had focused the attention of the public toward the inadequacies of the present calendar and the need for change.

The 13-month calendar was ably promoted in the United States by its supporters who took great satisfaction in emphasizing the number 13 as being particularly significant to America. They said that America was founded by 13 States, its flag had 13 stripes, and the Constitutional Seal abounded in many 13 motives. "Thirteen is not unlucky," they declared, admonishing Americans not to worry about the many Friday thirteens that occurred in this plan. They attempted to counteract the superstition by stressing that 13 was America's lucky number.

The progress of the 13-month calendar did not run smoothly, however, notwithstanding the ability, forcefulness and resources that went with it. Enthusiastically accepted by many business and statistical men, the public at large hesitated, more notably perhaps in Europe than in the United States. The sponsors met with formidable obstacles. Not only did tradition oppose, but mathematically the number 13 was a difficult one with which to cope. The convenient equal and half-year divisions by months were no longer possible.

The 13th month "Sol," placed between June and July, also gave rise to apprehension, in that an additional month would increase labor, expense, time and effort. To most Americans the mere thought of celebrating their national Independence Day on Sol 17 was repellent. The economic and statistical emphasis, too, was objectionable to many who felt that the use of the calendar covered a wider field and should not be so restricted.

Then, there were those who had a superstitious aversion to Friday 13. Once again let us quote from H. G. Wells'

*Outline of History* wherein he writes significantly on the aversion of the number 13 which he said was natural:

"And Neolithic man was counting, and falling under the spell of numbers.... He was beginning to use tallies, and wondering at the triangularity of three, and the squareness of four, and why some quantities like twelve were easy to divide in all sorts of ways, and others, like thirteen, impossible. Twelve became a noble, generous, and familiar number to him, and thirteen rather an outcast and disreputable one." [19]

An interesting comment is found in the report of the League of Nations on calendar revision in the footnote to Mr. Delaporte's 13-month plan which had been submitted to the League (date not given). It reads:

"This as an auxiliary system does not seem directly to concern calendar revision."

And that is just what it proved to be. There are certain business organizations today that use accounting methods on a 13-period "auxiliary system" in connection with our changeable Gregorian calendar. It seemed the only way by which some sort of comparability could be had with the past, present and future records. It was a makeshift. But to subject the calendar and all its users to such an arbitrary system would have been unjust and followed by chaotic consequences. It would have destroyed the rhythm of the calendar with its 12-month divisions, and the balance and equity between half and quarter-years by months would have been impossible. Months, too, would have been reduced to regimented four-week periods monotonous in their arrange-

[19] New and revised edition, Garden City Publishing Co., p. 123.

ment; historical records and accurate comparability with the past and present would have been lost; and the *337 conversion of dates* from the old to the new would have imposed a gigantic task.

The 13-month plan in its cyclonic change proved its own undoing. It dealt in extremes and nature ever resents extremes. The late poet laureate, Robert Bridges, in his *Testament of Beauty*, wrote:

"Our stability is but balance."

The fundamental law of balance was not recognized by the sponsors of the 13-month calendar. And yet by informing the public of the drawbacks of the present calendar, the pioneer work of the advocates of the 13-month calendar, particularly that of Mr. Cotsworth,[20] was most valuable. They did the hard spade work, plowing and making ready the way for a more balanced and equalized plan. The middle-of-the-way revisions are usually the ones that win through, endure and lead civilization onward to greater perfection.

[20] In the process of publication, word was received that Mr. Cotsworth died at Vancouver, B. C., in the evening of June 4, 1943, at the age of 83.

# 10

# LEAGUE OF NATIONS AND CALENDAR REVISION

Before we enter upon the middle-of-the-way revision of the calendar a brief review of the League of Nations' activities in dealing with the calendar is desirable.

The greatest impetus for an improved time-plan came from business through the International Congress of Chambers of Commerce and Industrial and Commercial organizations, which at its last three biennial meetings in 1910, 1912 and 1914, favored calendar revision and worked for an international conference on the subject. Members of the International Congress were constantly experiencing economic handicaps, trade difficulties and social inconveniences with the use of our ever changeable and irregular calendar. The Congress at its final meeting in 1914 requested the Swiss Government to make a careful study and analysis of the entire matter and to report on the question, preliminary to the holding of an international conference. Then came World War I and the revision was laid aside. Nothing further was done about it until the League of Nations was set up in Geneva. At the first meeting of the newly organized International Chamber of Commerce in 1920, successor to the former International Congress, it requested the League to consider the entire matter in view of finding a solution to the problem of obtaining the best calendar and in reaching an agreement on an international conference.

The League of Nations accepted in 1923 and through its Advisory and Technical Committee for Communications and Transit, a Special Committee was formed of which the then president of the International Chamber of Commerce, Willis H. Booth of New York, became a member. The others were W. J. M. van Eysinga of Holland, professor at the University of Leyden, chairman, appointed by the League of Nations; Professor M. G. Bigourdan, former chairman International Astronomical Union's Committee on the Calendar, appointed by the League of Nations; Professor D. Eginitis of Greece, director of the Observatory of Athens, appointed by the Ecumenical Patriarch of Constantinople; the Reverend T. E. R. Phillips, secretary of the Royal Astronomical Society of London, appointed by the Archbishop of Canterbury; and Reverend Father Gianfranceschi, president of the Academy "Dei nuovi Lincei," appointed by the Holy See. It was significant that the three religious groups in Europe—Anglican, Eastern Orthodox and Roman Catholic— reported that there are no dogmatic objections to the revision of the calendar. As to a fixed Easter, however, it is natural that as this is essentially a religious feast day the matter should rest with the churches.

After several meetings held by the Special Committee of the League of Nations, it was decided in 1926 to dissolve, at the same time requesting the League to advise its Member States and non-Member States to form their own calendar committees. This was done in order to prepare for an international conference in 1931. It was hoped that some agreement would be reached by which the new calendar could be put into operation by Sunday, January 1, 1933. This was a most appropriate date, as on that day, month and year both the old Gregorian and the new calendar would automatically meet, and the new calendar glide into the daily activities without disruption. The stage was being set.

In the discussions and meetings held at Geneva it was unfortunate that the League of Nations did not envision clearly the distinction between a revised civil calendar and that of fixed religious days, such as Easter. Great Britain, owing to her northern latitude and strong religious feeling for Easter, had passed the Parliamentary Easter Act in August 1928, placing Easter on "the first Sunday after the second Saturday in April." This was only to be operative with the approval of other Christian Churches. Because of our changeable calendar it was necessary to define the fixed Easter in so wordy a manner. Conciseness and simplicity were not possible. This British Act is at present a non-operative statute law. We shall see that it exerted a strong influence at the forthcoming international conference of the League of Nations.

The guiding spirit in accomplishing this "Act" was Lord Desborough, who succeeded in awakening the Anglican State Church to the reasonableness and value of observing Easter on a fixed Sunday just as Christmas is observed on a fixed date. Lasting tribute is due him. It is impossible, however, to observe a fixed day jointly with a fixed date in the present calendar. This can be achieved only in a perpetual calendar, one that is every year the same, unalterable in its regular and orderly arrangement.

## INTERNATIONAL CONFERENCES

At Geneva, from June 8-13, 1931, the Preparatory Conference held its meeting and from October 12-19, the International Conference on Calendar Reform deliberated on the subject. For interesting detailed information and inside activities of the conferences and of the eight years preceding them, I refer my readers to the *Journal of Calendar Reform* for 1942, wherein Mrs. Key-Rasmussen, of the Secretariat of the League of Nations, writes historically on the entire subject and proceedings.

At both conferences the League had invited representatives of various calendar proposals, as well as religious groups who might approve or oppose calendar revision, to present their views. The public hearings were held on the first days of each assembly, the speakers given from ten to fifteen minutes, after which the conferences were closed to all but official delegates. Besides the well-known, and in many quarters what was believed to be the successful 13-month plan, a newcomer had appeared in the form of The World Calendar of 12 months and equal quarters. It proved a strong contestant at both conferences, successfully stalemating the 13-month proposal, so that the International Conference closed with no tangible result. Two nations, Canada and Yugoslavia, had approved the 13-month plan; and two, Switzerland and Greece, the new perpetual World Calendar of 12 months and equal quarters. The whole question was resubmitted by the League to the various governments for further study with the request that wider dissemination be given the public on the matter. But real good had resulted. Out of more than 500 plans submitted to the League only these two survived for further consideration. This in itself was distinctly worth while.

Among the many rejected plans the most notable were:

*First,* those advocating four long 35-day months and eight short 28-day months, rejected "because the very perceptible inequality of the months would be extremely inconvenient from every point of view";

*Second,* plans proposing a "Leap-Week Calendar" with an extra leap week every five, six, or even eleven years, rejected on the ground that they are "inferior to the existing calendar and cannot be considered at all";

*Third,* plans for a decimal system of time-measurement;

*Fourth,* plans for a five, six or ten-day week;

*Fifth,* plans for changing the date of the New Year;

*Sixth,* plans proposing a mere equalization of the year by transferring the last day of August to the last day of February.

Strong opposition to calendar revision from certain religious groups had developed within the two conferences, especially among the Orthodox Jews and the Seventh-Day Adventists. The silence of the Vatican was also interpreted by some as a sign of opposition.

The League, however, following Great Britain's lead, passed an Easter Act establishing Easter on "the first Sunday after the second Saturday in April." [21]

The strange anomaly of a secular organization like the League of Nations dealing with a fixed Easter, although meticulously stressing the economic and social aspects, had estranged in some measure certain churches. And the Orthodox Jewish and Seventh-Day Adventists' opposition to the supplementary or extra days, as breaking into the al-

[21] See Chapter XVII, pp. 115-116.

leged unbroken sequences of seven-day weeks, also exerted an influence against the revision.

As an official observer at the League of Nations, it was my privilege to listen to the impassioned arguments of the opponents. It made one yearn for a Moses, a Confucius, a Buddha, a Jesus, to strike from man the shackles of his own making and free him from his man-made traditions. It seemed so unreasonable for the past to exert such a stranglehold that it should hinder his progress toward growth and advancement, toward better days to come.

Once again, as in the past, enslaving tradition and custom interfered. The universality of the calendar as a time-system belonging to everybody and to the whole world was not understood. Individualism and partisanship were not ready to accede to greater co-operation and universality. Here was a clear indication that future work had yet to be done for the revision, further enlightenment had yet to be attained before a new world calendar would be accepted in a spirit of greater world amity and understanding. Group selfishness, group fear, group superiority and group power were still too great, and the League of Nations too ineffective to accomplish its objective. Although the League failed in its attempt, it did succeed in placing calendar revision clearly and squarely before civilization for further action.

## Later Activity of the League

With the international conference concluded, the two calendar protagonists continued their work with renewed vigor, undaunted by the temporary disappointing result. It seems an inevitable truism that revisions to prove their merit must meet with adverse as well as with favorable reception.

Only by such tests do they survive and succeed or weaken and disappear.

Of the two plans the rapid decline of the 13-month calendar was a surprise. More and more its drawbacks became apparent, and it could not overcome the lamentable death of its supporter, George Eastman, in 1932. It was about 1936 that both committees, the one in London and the other in Rochester, quietly closed their doors and further activities ceased. Certain organizations still use a 13-period accounting system in connection with the present Gregorian calendar, but as a new and better calendar for world-usage it had failed.

In contrast, the other plan, The World Calendar of 12 months and equal quarters, has gained steadily in strength and support. Science, commerce, education, government, and organizations of men and women, all are being attracted to it on the sheer merit of the plan. It is justifiable to state that it is the plan itself that is winning friends wherever it is studied, wherever it is known, rather than any one personality or group. The World Calendar is truly global and invites global co-operation and global recognition.

Thus it was that the Labor Conference of American States in Chile, 1936, passed a strong resolution in its favor. This induced Chile to give The World Calendar careful study and consideration. Through the guiding spirit of the late Dr. Ismael Gajardo Reyes, chairman of the Chilean Calendar Committee, a resolution was drafted and presented to the League of Nations requesting that the League Council submit this draft to the various governments for their opinion and approval. It was confidently and justifiably believed that the answers would pave the way for another international conference in the autumn of 1937. The new calendar could then have been put into operation in 1939, the year when again both the old and the new calendars would meet on the same day and date—Sunday, January 1—and the transition be an easy one.

Under the notable sponsorship of the late Ambassador H. E. Augustin Edwards, who held the Chilean portfolio in Great Britain and was a member of the League Council, the League Council accepted the Chilean draft resolution for submittal to the various governments, March 1937. Among the 45 replies received:

*Six* governments rejected calendar reform—Argentina, Haiti, India, Netherlands, Roumania, Sweden;

*Fourteen* accepted The World Calendar draft—Afghanistan, Brazil, Chile, China, Spain, Esthonia, Greece, Hungary, Mexico, Norway, Panama, Peru, Uruguay, Turkey;

*Eight* made no observations—Australia, Denmark, Iraq, Iceland, Liechtenstein, Siam, United States of America, Venezuela;

*Ten* were unprepared—Bulgaria, Egypt, Ecuador, Finland, Guatemala, Latvia, Lithuania, Switzerland, Czechoslovakia, Union of Soviet Socialist Republics;

*Seven* considered the time premature—South Africa, Austria, United Kingdom, France, Monaco, New Zealand, Poland.

As mentioned above the reply of the United States was noncommital.

Of the fourteen nations who officially approved the new calendar there were seven of the Roman Catholic belief; three Protestant; two Mohammedan; one Eastern Orthodox; and one Taoist and Buddhist—China. This was encouraging. It demonstrated in a remarkable degree the universality of the civil calendar, and that a revised calendar could overcome religious, racial and national biases, uniting them by a common bond of time.

The fourteen nations formed the nucleus on which to build further. An unusual opportunity had presented itself for world agreement on one definite objective—the calendar. It was highly regrettable that the League and its members did not recognize this far-reaching significance.

Did Charles Nordmann, eminent French astronomer, in his book, *The Tyranny of Time,* published in 1925, have a foreboding when he wrote?

"The fate of this question [calendar reform] will enable us to judge the League of Nations.

"Some persons believe the League is the chrysalis whence a better future will emerge on the luminous wings of Justice and Liberty. Others hold it is but the last and fugitive reflection of the illusory clouds of the Past. But belief is not knowledge. The reform of the calendar will help us in choosing between the two points of view." [22]

Had the members of the League read and studied this book, the result might have been different.

The attitude of the Vatican toward the Chilean draft also contributed to uncertainty and confusion. It was natural to assume that as its own Abbé Mastrofini, an Italian priest, had conceived the perpetual calendar containing the one or two supplementary days, and that The World Calendar had the support of Chile together with other Roman Catholic countries, the Vatican would have supported it. But the Vatican remained aloof. Perhaps it realized more clearly than did the League that a civil calendar revision actually

[22] International Publishers, New York, 1925, p. 100.

belongs to the jurisdiction of governments for decision, whereas the fixed Easter date belongs to religious authorities. We need only to recall the Papal declaration of 1912 during the incumbency of Pope Pius X.

As a result of an inquiry sent to the Vatican by the International Congress of Chambers of Commerce at their meeting held in Boston in 1912, the following reply was received:

"The Holy See declared that it made no objection but invited the civil powers to enter into an accord on the reform of the civil calendar, after which it would willingly grant its collaboration in so far as the matter affected religious feasts." [23]

It was unfortunate that the League of Nations was either not aware of this statement, or if it were, that it was not more closely followed. Be that as it may, the League of Nations once again put aside calendar revision. And this notwithstanding the valiant efforts of H. E. Alfredo de Castro, Uruguayan Minister at Belgium and member of the League Committee on Communications and Transit.

The League was too weak, however, to lay a foundation upon which its prestige would have been enduring—an unforgettable record of a great world-wide and international achievement—the sponsoring and establishing of a new and better world calendar.

[23] *The Catholic News,* March 27, 1937; *Journal of Calendar Reform,* Volume 7, p. 85.

# 11

# "THE VERY BEST"

Now we approach the new and better revision of the civil calendar which is ready, and the support it is receiving justifies the belief that it will be adopted by 1945.

"A new cycle of time is in its inception." [24]

In the summer of 1929, I was vacationing with friends at the Lake Placid Club in the Adirondack Mountains. Hearing of a lecture on "How to Simplify Life," I decided to attend as the subject was intriguing. I invited some of my friends to join me but as they all refused I went alone to hear the views of Dr. Melvil Dewey, famous for his filing system in our public libraries, and founder of the club. Dr. Dewey divided the lecture into three subjects: the first, simplified spelling; the second, the need of a universal weight and measure system; and the third, the importance of improving the calendar. The third aroused my interest as I had never given the calendar particular thought, taking it for granted. Its familiarity had impressed itself as enduring, and now I was learning that it had been changed before and could be changed again. I was also told of its many defects and how

[24] Fullwood, Nancy, *The Song of Sano Tarot*, p. 50, Robert M. McBride & Co., New York, 1930, by permission of her son, Harry Fullwood.

these were really handicapping our activities—our very lives in fact. I was becoming absorbingly interested and keenly awaited the new plan. It came as a 13-month proposal and one that I believed originated with Dr. Dewey. As he elaborated I became increasingly indignant. How could the prime number 13 ever simplify? How could such an arbitrary and indivisible number ever replace the convenient 12? Something deep within me stirred in protest so that I left the lecture feeling decidedly uneasy. My friends laughed at my discomfiture and told me to forget it. And that is what I did, for what could I do otherwise?

It was about a fortnight later, on Sunday morning, September 8, that I was leisurely reading *The New York Times*. Quite unexpectedly I saw a letter to the Editor wherein the writer, Lewis E. Ashbaugh of Denver, Colorado, wrote that a year divided into quarters is a custom used in Europe far more than in America, and concluded:

"While we are planning an improved calendar, let us also insist on the very best, with all conditions considered, and let us adopt the revised twelve month year of equal quarters and equal working-day months, easily adapted from the calendar we now use."

A detailed plan which accompanied the letter was so beautifully symmetrical in arrangement, so ordered and simple with its equal quarters, that I felt this to be the right one. Then and there I decided to work for it, and from this decision I have never wavered. Like Mr. Cotsworth, I found my life-work and felt privileged in devoting my time and effort toward its adoption. For what could be more constructive than to help the world in obtaining a more ordered, simple and stabilized calendar? One which would serve as a steadying force, uniting all nations in a better measurement of time?

Upon my return home, I arranged my affairs so that I could devote myself unhampered to this new work, and in 1930 began earnestly to study the whole question. The New York Public Library was an admirable source of information, and yet, I was dismayed to find such meager material on the calendar. I gradually learned that scientific pamphlets, treatises and magazines were frequently more informative, but also more difficult to obtain.

In the autumn with the aid of friends, The World Calendar Association was incorporated and the first pamphlet *The World Calendar* was widely circulated. It was mailed immediately to the League of Nations and the two calendar committees which sponsored the 13-month plan. Editors of the press and specialized lists were also included.

The newly organized World Calendar Association was incorporated October 21, 1930. As the Preparatory Committee of the League of Nations was scheduled for June 1931, something had to be done immediately to place the new time-plan in the forefront of action and to bring it to the attention of the League. Time was of necessity short, every minute of the intervening eight months was precious.

The Association, through its first director, Carl Liddle, was fortunate in securing the services of Charles Dexter Morris, who, as a former reporter of the Associated Press, publicity director of the American Red Cross in Paris, and an agent of the Near East Foundation, was well versed in European conditions and the League of Nations. It was his idea that the Association publish a quarterly *Journal of Calendar Reform,* scientific in approach and representative in appearance, so that it would instantly impress the League with an authority that could not fail to be impressive. The first publication proved a valuable talisman and the continued quarterlies are an open sesame bringing friends and endorsers to The World Calendar from every corner of the globe.

# THE WORLD CALENDAR

This 12-month-equal-quarter calendar is the same for every year.

The quarters are equal in length.

Each quarter begins on Sunday and ends on Saturday, contains 3 months—13 weeks—91 days.

Month-dates always fall on the same weekdays. Each month has 26 weekdays—plus Sundays.

Year-End Day and Leap-Year Day, the extra Saturdays, December W and June W, are World Holidays.

Each year begins on Sunday, January 1, and the business year begins with Monday, January 2. Because a World Holiday precedes Sunday, the usual custom of celebrating a Sunday holiday on Monday is voided.

This revised calendar is balanced in structure, perpetual in form, harmonious in arrangement.

# THE WORLD CALENDAR

### EACH YEAR THE SAME

## FIRST QUARTER

| | S | M | T | W | T | F | S |
|---|---|---|---|---|---|---|---|
| JAN | 1 | 2 | 3 | 4 | 5 | 6 | 7 |
| | 8 | 9 | 10 | 11 | 12 | 13 | 14 |
| | 15 | 16 | 17 | 18 | 19 | 20 | 21 |
| | 22 | 23 | 24 | 25 | 26 | 27 | 28 |
| | 29 | 30 | 31 | | | | |
| FEB | | | | 1 | 2 | 3 | 4 |
| | 5 | 6 | 7 | 8 | 9 | 10 | 11 |
| | 12 | 13 | 14 | 15 | 16 | 17 | 18 |
| | 19 | 20 | 21 | 22 | 23 | 24 | 25 |
| | 26 | 27 | 28 | 29 | 30 | | |
| MAR | | | | | | 1 | 2 |
| | 3 | 4 | 5 | 6 | 7 | 8 | 9 |
| | 10 | 11 | 12 | 13 | 14 | 15 | 16 |
| | 17 | 18 | 19 | 20 | 21 | 22 | 23 |
| | 24 | 25 | 26 | 27 | 28 | 29 | 30 |

91 days

## THIRD QUARTER

| | S | M | T | W | T | F | S |
|---|---|---|---|---|---|---|---|
| JUL | 1 | 2 | 3 | 4 | 5 | 6 | 7 |
| | 8 | 9 | 10 | 11 | 12 | 13 | 14 |
| | 15 | 16 | 17 | 18 | 19 | 20 | 21 |
| | 22 | 23 | 24 | 25 | 26 | 27 | 28 |
| | 29 | 30 | 31 | | | | |
| AUG | | | | 1 | 2 | 3 | 4 |
| | 5 | 6 | 7 | 8 | 9 | 10 | 11 |
| | 12 | 13 | 14 | 15 | 16 | 17 | 18 |
| | 19 | 20 | 21 | 22 | 23 | 24 | 25 |
| | 26 | 27 | 28 | 29 | 30 | | |
| SEP | | | | | | 1 | 2 |
| | 3 | 4 | 5 | 6 | 7 | 8 | 9 |
| | 10 | 11 | 12 | 13 | 14 | 15 | 16 |
| | 17 | 18 | 19 | 20 | 21 | 22 | 23 |
| | 24 | 25 | 26 | 27 | 28 | 29 | 30 |

91 days

## SECOND QUARTER

| | S | M | T | W | T | F | S |
|---|---|---|---|---|---|---|---|
| APR | 1 | 2 | 3 | 4 | 5 | 6 | 7 |
| | 8 | 9 | 10 | 11 | 12 | 13 | 14 |
| | 15 | 16 | 17 | 18 | 19 | 20 | 21 |
| | 22 | 23 | 24 | 25 | 26 | 27 | 28 |
| | 29 | 30 | 31 | | | | |
| MAY | | | | 1 | 2 | 3 | 4 |
| | 5 | 6 | 7 | 8 | 9 | 10 | 11 |
| | 12 | 13 | 14 | 15 | 16 | 17 | 18 |
| | 19 | 20 | 21 | 22 | 23 | 24 | 25 |
| | 26 | 27 | 28 | 29 | 30 | | |
| JUN | | | | | | 1 | 2 |
| | 3 | 4 | 5 | 6 | 7 | 8 | 9 |
| | 10 | 11 | 12 | 13 | 14 | 15 | 16 |
| | 17 | 18 | 19 | 20 | 21 | 22 | 23 |
| | 24 | 25 | 26 | 27 | 28 | 29 | 30** |

91 days

## FOURTH QUARTER

| | S | M | T | W | T | F | S |
|---|---|---|---|---|---|---|---|
| OCT | 1 | 2 | 3 | 4 | 5 | 6 | 7 |
| | 8 | 9 | 10 | 11 | 12 | 13 | 14 |
| | 15 | 16 | 17 | 18 | 19 | 20 | 21 |
| | 22 | 23 | 24 | 25 | 26 | 27 | 28 |
| | 29 | 30 | 31 | | | | |
| NOV | | | | 1 | 2 | 3 | 4 |
| | 5 | 6 | 7 | 8 | 9 | 10 | 11 |
| | 12 | 13 | 14 | 15 | 16 | 17 | 18 |
| | 19 | 20 | 21 | 22 | 23 | 24 | 25 |
| | 26 | 27 | 28 | 29 | 30 | | |
| DEC | | | | | | 1 | 2 |
| | 3 | 4 | 5 | 6 | 7 | 8 | 9 |
| | 10 | 11 | 12 | 13 | 14 | 15 | 16 |
| | 17 | 18 | 19 | 20 | 21 | 22 | 23 |
| | 24 | 25 | 26 | 27 | 28 | 29 | 30* |

91 days

\* A WORLD HOLIDAY, DECEMBER W, the Year-End Day, an extra Saturday, follows December 30 every year.

\*\* A WORLD HOLIDAY, JUNE W, the Leap-Year Day, another extra Saturday, follows June 30 in leap years.

# 12

# THE WORLD CALENDAR

In planning the new calendar which follows the middle-of-the-way revision, one thought rises above all others. It must be world-wide: world-wide in application, world-wide in appeal and world-wide in scope. The new and better calendar cannot be planned primarily for the English speaking peoples, the Latin, the Teutonic, the Slav or the Asiatic. It must avoid religious labels such as Christian, Jewish, Mohammedan or Hindu. Its usefulness must not be restricted to economic, trade, and governmental fields to the exclusion of the educational, scientific, cultural, social and religious. Patriotism, partisanship and politics have no privilege in its planning. This is not difficult of accomplishment if everyone is willing to sacrifice, to give, to share and to accept, for the greater good of all mankind. In the realm of Time, different groups merge and become as one. The new calendar must be "the very best, with all conditions considered."

In this connection, I wish to tell of an incident previous to the last World War that made a great impression upon me. I had anticipated reading with real enjoyment Houston Chamberlain's book *The Foundations of the Nineteenth Cen-*

*tury.* But as I progressed I became increasingly aware of the writer's bias and concentration on the "Nordic race." He believed it was the superior race upon which rested the culture and future of civilization. I finished the book very much disturbed with the blind arrogance and limited vision of the author. Instantly I recalled the words of Jesus, "In my Father's house are many mansions." And are there not many nations and peoples in the world, I reasoned? This was my first conscious awareness of the erroneous belief in the superiority of any one race, any one nation, any one belief. This experience, I realize now, was a necessary foundation for my work for a better calendar for the use of *all* mankind.

The World Calendar offers "the very best," for it not only retains the familiar 12 months but it achieves perfect equality in its quarter and half-year divisions. Every quarter begins on a Sunday, ends on a Saturday, and has 3 months, 13 weeks or 91 days. The 3 months are more evenly apportioned by giving the first month 31 days and the other two 30 days each, and every month has 26 weekdays plus Sundays. In this manner the calendar avoids a rigid monotony that stultifies, and brings to it an ordered variety that enlivens. Every year begins on Sunday, January 1; every year is comparable to every other year; and what is of utmost importance, *days and dates always agree.* The civil calendar has at last found its focal point and so stands proudly in its own right beside the compass and the clock.

The 365th day necessary to complete the year and the 366th day in leap years, hitherto called supplementary days, are now known as the two new WORLD HOLIDAYS. They give to the calendar the much needed stability wherein every year is like every other year. Each year begins on Sunday,

January 1, and closes with the Year-End Day, December W, a World Holiday, placed on an extra Saturday.

The 366th day in leap years is the new Leap-Year Day, placed in mid-year, thereby keeping the year balanced and equalizing the half-year divisions. It is another World Holiday, extra Saturday, June W. In this simple arrangement each calendar-unit fits perfectly within the quarter, bringing a co-ordination never known before.

The introduction of the one or two extra days, the new World Holidays, suggests a delightful Chinese story about a farmer who at his death left eleven sheep to his three sons, with the following bequest: that to his oldest be given one-half the number of sheep, to the second one-quarter, and to the youngest two-thirds of the remainder.

This strange division greatly distressed the sons, until a distinguished mathematician showed them a way of solving their problem. He told them to go to a neighbor and ask for the loan of one sheep. With this borrowed sheep the sons now had twelve animals, which they could distribute in accordance with the wish of their venerable father. The oldest son received one-half the number, or six sheep; the second son received one-quarter, or three sheep; the youngest received two-thirds of the remainder, or two sheep. When the borrowed animal had thus performed its duty, the sons returned it to their neighbor.

A similar problem confronted calendar students intent upon improving the calendar. How could the 365-day year be made stable and divisible? They adopted a solution just the reverse. Instead of borrowing a day they temporarily removed a day, by which method the calendar became an equally divisible 364-day year. This accomplished, the removed

365th day, so important to the seasonal and scientific accuracy of the year, was returned to the calendar and placed at the end of the year as the extra Saturday or World Holiday. Thus the calendar became complete and changeless, and the accuracy of the calendar year was maintained.

Referring once again to the significance claimed for the number 13 in reference to the United States, it could be said that the 13 weeks within the equal quarter divisions of The World Calendar also commemorate the 13 original States which gave birth to the nation. But the new time-plan goes beyond this early restriction. It extends to the entire country as it is today—composed of 48 States. While 13 was a beginner, it was not a consummator of the nation.

The 48 States, symbolized in the 48 stars of the flag, are the sum total of 4 x 12, suggestive of a square. The United States is based on a solid square which the national seal portrays in the symbol of the four equal-sided pyramid, and which has been given further national expression in the majestic shaft of sheer dignity and simple beauty—the Washington Monument, also based on the square. Thus the number 12 seems to have more significance to America than 13.

The World Calendar is a mathematical masterpiece. It deals as successfully with the difficult numbers 7 and 13 as with the easy numbers 2, 3, 4 and 6. And by means of the new stabilizing World Holidays the calendar has become a steady time-system.

It is remarkable that there is no one person after whom the calendar can be named. The World Calendar gradually evolved from Switzerland, first, as the result of a study made at the request of the International Congress of Chambers of Commerce in 1914, and later through the activities of the

League of Nations. Many minds have contributed, valuable historical facts have been unearthed, greater knowledge and understanding have been gained, and yet no one can claim it.

As the name implies, the new calendar belongs to the world: it is world-wide in purpose, usage, and possession.

"Time that 'takes survey of all the world' should itself be one and the same for all the world." [25]

[25] *The New York Times,* an editorial "Calendar Contentions," August 26, 1935.

# 13

# THE WORLD HOLIDAYS

THE TWO STABILIZING DAYS of the new time-plan call for special attention. They were conceived as we have seen by an Italian priest, Abbé Mastrofini, in 1834, and named blank days, set apart as outside the 364-day year. These "blank" days met with serious objection, however, because in the literal sense there can be no blank days in the calendar or in our lives. Every day lived is a living day and cannot be thoughtlessly slighted or nullified. The terminology "blank" was therefore most unfortunate, inviting opposition.

It was not until the League of Nations undertook the study of calendar revision that these troublesome days were re-named intercalary or supplementary days. They still re-mained outcast calendar-children, however, without a day of the week and a date of the month to call their own. The new scientific names were a decided improvement, but they were cold in their abstract and intellectual isolation. They carried no appeal. It was not until The World Calendar Association gave these days their joint name, World Holi-days, with their individual names—Year-End Day and Leap-Year Day—placing them on extra Saturdays with their own particular dates—December Y or 31 and June L or 31—that the former opposition lessened.

The extra Saturday idea is a revival of an ancient practice, when occasional feast days were observed for 48 hours. Rabbi Martin M. Weitz tells us that according to early Hebrew and Babylonian records:

"If it [The World Calendar] sanctified the week additionally in that it can reintroduce an ancient Jewish practice—a *48 instead of 24-hour* 'coverage' for major rest-days and festivals, it may well be time to lengthen again the one-day and one-week festivals by one day for each. Thus Passover would be celebrated by *all* for eight, not seven days, and New Year's, for two, not one days. Orthodox Jewry celebrates eight and not seven days for Tabernacles and Passover, and two, not one, for New Year and Pentecost, in order that Jewry all over the world shall be able to celebrate these festivals simultaneously." [26]

In giving these days a name, a weekday, a date and a month, their purpose became better understood. It was the only method by which the calendar could become stabilized without losing its well-won scientific accuracy in the seasonal year.

And yet something was lacking. The World Holidays needed a universal denominator. The Year-End Day, December Y, differed too greatly in countries where the last day of the year is called New Year's Eve or Sylvestre. And Leap-Year Day, June L, did not serve either, as this day is Bissextile in many countries.

The recent improvement to call the 365th and 366th days World Holidays, dating them December W and June W, eliminates the previous awkward tabulation. Thus the letter W denotes the purpose and place of the one or two World Holidays. Every year closes its cycle on the extra Saturday World Holiday, December W, and every leap year is augmented in the mid-year by the extra Saturday World Holiday, June W.

The letter W is also remarkably applicable in a universal sense. In the various root languages the letter W stands for world, welt, etc., and reversely the letter M stands for Mundo, Monde, etc. The letter W is also suggestive of two Vs, denoting a double victory—a victory for stability and a victory for world observance. Without doubt these days observed throughout the world will foster a greater world unity, co-operation and good will in this new and better measurement of Time. Are they not truly symbolic of the leaves for the *healing of the nations,* so beautifully described in the promise made in *Revelation* where it tells of the tree of life?

"The tree of life, which bare twelve manner of fruits, and yielded her fruit every month: and the leaves of the tree were for the healing of the nations." [27]

Notwithstanding that the one or two extra Saturday World Holidays are essential to the calendar if it is to be stabilized and if it is to maintain its accurate astronomical length every year, opposition still exists. The opposition rests on the premise that these days would create an eight-day week at the close of every year and another in the middle of the calendar in leap years, thereby breaking the alleged continuity of the seven-day week since "time immemorial."

[27] *Revelation* 22:2.

This assumption of the unbroken seven-day week which history, as we have seen, does not substantiate [28] appears the paramount issue among certain orthodox groups. And thus no matter to what religious sect these groups belong, any change that violates this alleged concept is, in their opinion, unacceptable.

Again man-made tradition and custom stand in the way of progress. The use of any civil calendar which achieves scientific accuracy and stability every year brings to those who hold this arbitrary concept a "wandering Sabbath."

We have just read that "Orthodox Jewry celebrate eight and not seven days for Tabernacles and Passover and two, not one, for New Year and Pentecost, in order that Jewry all over the world shall be able to celebrate these festivals simultaneously." The one eight-day week every year and another in leap years does not appear to be at variance with this principle.

To quote Dr. Julian Morgenstern, president of The Union Hebrew College at Cincinnati:

"In principle, I am in sympathy with the project of The World Calendar Association, particularly, if it is not at all the purpose of the Association to interfere unnecessarily in the religious calendars and ceremonial observances of various sects. I believe that it would be expedient for the Association to emphasize this fact over and over again. . . .

"With regard to the traditional Jewish religious calendar, the great difficulty in the way of correlating The World Calendar with it lies in the fact that The World Calendar would throw the traditional Jewish Sabbath out of place, in six years out of every cycle of seven years. I fear that this is an insurmountable

[28] See Chapter 6, pp. 36-37.

obstacle for Orthodox Jews. I have given thought to the matter and see no way in which this obstacle can be surmounted. Otherwise, there is no difficulty whatever in the recognition and employment of The World Calendar by Jews for civil purposes.

"Of course, should our Government ever officially recognize the civil World Calendar, American Jews would accept this calendar readily and apply it for civil purposes. It would then become their responsibility to find a way to harmonize their traditional religious observances with the new and now official calendar.

"They have faced this problem before and found a solution. And I have no doubt that they could do it again if the need arises." [29]

Is not this attitude reasonable and will it not remove former opposition to the new civil calendar?

The outstanding value to be had from the perpetual World Calendar is that every new year will always begin on Sunday, January 1, the first day of the week. With the year always beginning on the day of the Lord and always ending with a Sabbath of rest, followed by the World Holiday, December W, may we not receive thereby untold spiritual blessings? All that moves forward together in unity and good will promises "one for all and all for one."

The World Holidays, each coming between the Jewish Sabbath and the Christian Sunday, are like the hands of God stretched out from ageless time to welcome both the Saturday and the Sunday of rest, prayer and worship in closer relationship. Like the rainbow bridge that arched the sky after the flood with the promise of better days to come, so the new calendar shines forth, after hundreds of years

[29] *Journal of Calendar Reform,* Volume 12, p. 128.

flooded with disorder and discord, and promises better days to come through the two new World Holidays. They form a rainbow bridge of many colors whereon peoples of all climes, customs and faiths join in a spirit of greater fellowship and friendship. THE WORLD HOLIDAYS ARE THE FRIENDLY HANDCLASPS OF TIME.

# 14

# THREE ANALOGIES

"The time is out of joint; O cursed spite,
That ever I was born to set it right!"

THIS WAS THE despairing and hopeless cry of a sensitive
Hamlet. But today no such thought obsesses us. With faith
and determination in better days to come we hopefully
acclaim:

Our time *is* out of joint; but O delight,
That we are born this age to set it right!

The new and better time-plan will make of our Earth a
more wholesome planet on which to live, since her confused
and irregular breathing of the past will have become quiet
and regular. And with the return to balanced and ordered
breathing, nations, governments, science, education, agriculture, business, culture, social life and religion, all our many
fields of activity, will be benefited. Days and dates are one in
agreement. The ordered months, the equal quarters and
half-years, and the regular number of weeks within every
quarter-year are the signs of Earth's restored health. She
has found a greater equilibrium.

It matters not whether one group functions by the day
system, another by the week, or month, or quarter, or season.
The different time-units all form one complete quarter-year,

which repeated, combine in forming one co-ordinated year, augmented by a new World Holiday (the Year-End Day, December W) that seals the annual cycle with its benediction. Like the basic four seasons of nature and the four-square foundation of the pyramid, The World Calendar stands square with time, an enduring time-measurer for man's daily needs.

Let us picture the contrast between the shifting and discordant old with the steady and harmonious new calendar by supposing that all the states of the United States of America were warring against each other as do the different time-units in the present calendar. Each state claims superiority, each state is self-centered, unwilling to share and to give in any co-operative spirit. The United States in this picture is a warring country with chaos, dissension and confusion rampant.

Now let us look at the other side of the picture. Harmony dwells within the states. Each state is free to develop to its best capacity and resources exerting all its effort toward progress without interfering or infringing upon any other state. Each state respects every other state, and is willing to co-operate in unison. The states live together not in a spirit of rivalry and competition, but in neighborliness and co-operation. E PLURIBUS UNUM has made the United States of America a strong, co-operative and united country without sacrificing the freedom and the rights of the individual states.

From these two analogies, let us now progress to the broader aspect of the world. We have been prone to regard the world as made up of separate and distinct groups or nations, all more or less unrelated. Each group contends with

the others, each strives to possess and to have more than
the others, each claims superiority and power. There is thus
engendered a continuous contest of rivalry and competition
among them, paralyzing world progress and human welfare. The world is a warring world, like the first picture of
the United States. It is selfish and combative.

The other more hopeful picture is a world that has become
united and whole, comprising many neighborly groups or
nations all of which are free in their own rights. Each group,
whether large or small, rich or poor, strong or weak, old
or young, of whatever color, race or creed, recognizes that
each is a necessary and vital part of the entire world. Each
nation has the freedom to develop to its best capacity neither
more nor less, each is respected for what it is, and each and
all willingly co-operate in forming one harmonious and
equitable world. Not one group or nation dominates, neither
is one group or nation subjected to another, which suggests
Helen Keller's inspiring phrase:

"No nation is wise enough to rule another." [30]

These two pictures—a harmonious United States and an
equitable Federation of Nations—portray freedom and
equality, self-respect and honor, neighborliness and good
will, the goal of civilization.

From considering a nation, then the world, let us now
consider the solar system and, for a moment, conceive it to
be warring. There is Mercury dissatisfied to remain within
its prescribed circuit, forcefully breaking away to interfere
with its neighbor planet Venus; or the Earth reaching out

[30] Keller, Helen, *Let Us Have Faith,* p. 65, copyright 1940, Doubleday,
Doran and Co., Inc., New York.

to join Mars in a false and overweening ambition to possess more than it has; or again Saturn encroaching upon the domain of Jupiter; all acting without law, each following blindly and greedily its own selfish will. Were this actually to happen a cataclysm too terrible to contemplate would result. Happily there exists no warring solar system.

On the contrary it is the most harmonious, ordered and equitable system known to man, who stands before it in contemplation, awe and reverence, silent before its beauty, law and order. It is a system wherein every planet functions freely in its own right and *in its freedom allows every other planet to be free also,* yet all are interrelated and interdependent. There exists no desire for interference or encroachment, no desire to rule or to dictate. It is a perfect example of freedom within the immutable law of the universe which was recognized by a former president of the United States, Calvin Coolidge, in his book *Price of Freedom:*

"The process of civilization consists of the discovery by man of the laws of the universe and of living in harmony with those laws." [31]

Mindful of these three analogies how does The World Calendar compare with them? Does it belong to the warring and contentious types or to the harmonious and equitable?

We observe that The World Calendar in principle and arrangement recognizes and respects the logical differences and functions of the various time-units. To each unit is given freedom and equality. The days always agree with their dates without outside interference or encroachment. Likewise the seven-day week is a distinct unit yet all of the 13

[31] Charles Scribner's Sons, New York, 1924.

weeks agree with the other units at the end of every quarter-year. Each month with its newly allotted and more equally apportioned days enjoys its independence with 26 week-days plus Sundays. Each of the quarters, year in and year out, is composed of an even 91 days or 13 weeks or 3 months or its approximate season, all willingly co-operating. The new time-plan has achieved perfect co-ordination among its many interrelated units without any loss of their individual freedom and function. There is not a single time-unit that dominates, all blend harmoniously to complete and form the whole. And with the World Holidays, the new civil calendar has gained stability and maintained the scientific accuracy of the solar year.

# 15

# ADVANTAGES TO ALL

With the new and better civil time-plan clearly stated, let us be more specific concerning the advantages it will bring to our daily affairs—to you and me and the other fellow. For a balanced calendar naturally goes hand in hand with idealism and practicality. Whatever our interests and activities, The World Calendar, being on the square with time, influences us all. Naturally we cannot give all the advantages in detail. Each person according to his own interests can discover for himself other benefits as they apply to him.

## Science

Science, which gave birth to the calendar, rejoices that the new time-plan retains the accurate length of the solar year and conforms to the seasons. The valuable work of centuries is not undermined but augmented and brought a step closer to perfection, through the equal quarter-divisions of the year and the stabilizing of the calendar. The new found order and comparability of The World Calendar appeals especially to scientists.

## Religion

And religion, too, is benefited by the civil calendar. Fixed feast, fast and holy days agree with their days and dates, thereby bringing a greater sense of tranquillity into their observances. Both clergy and congregation gain with this new and better calendar.

## Industry

To the industrial world the orderly arrangement of the perpetual calendar is of the utmost importance. Business corporations are usually subdivided into several departments requiring different methods of payment. One department deals with the day or hour worker, so that the pay is based on the day. Another department conducts its financial operations with the week as a pay unit. Salaries are frequently paid on a semi-monthly basis, whereas the shipping, transportation, delivery or billing department uses the month for its payments. The financing of the corporation itself, that includes dividends, interest, taxes, and financial statements, is computed on a quarter-year basis. These various time-units for payment bring no confusion as they all agree at the end of every quarter-year, and the new World Holiday, Year-End Day, is considered a national holiday. The wheels of industry run smoothly with this co-ordinated time-plan.

## Home

This calls to mind the natural interest the new World Calendar will have in the home, for the homemaker has to budget her time and her work, food and money for her family's needs. The steady World Calendar is a veritable

boon, permitting her to care for her family and plan social welfare, civil and church activities with greater ease
and security. The planning of school-time and vacations for
her children will be also materially helped with this rhythmic
and perpetual calendar. Greater harmony in the family will
be possible because the unhappy family of time, as we knew
it in the Gregorian calendar, has been transformed into a
happy family of time in The World Calendar.

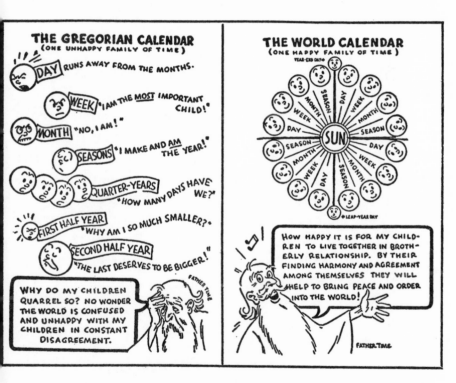

## Labor

For labor, the new time-plan is fair and just to both em-
ployer and worker. Whatever the time-period on which the
worker is paid, the payments always come at the same inter-

vals regularly—on the same weekday and correspondingly on the same month-date every year. This makes for more accurate planning and better budgeting in the home. Wages and expenses can be figured more easily and exactly whereby waste is prevented and saving encouraged. The World Calendar is a perfect timing-tool which is bound to be a valuable asset and a profit. For good tools always bring good results. And the new World Holidays, for labor as well as industry, will be treated as national holidays.

## Agriculture

To the agriculturalist, no matter what his product, be it grain, cattle, milk, vegetable or fruit, better results and more profits come with careful supervision and planning. Farm production is now "big business" and calls for definite comparability month by month, season by season, year by year, if crop rotations are to be planned, breeding records to be kept, milk and egg production to be measured.

The new civil calendar—conforming to natural order and regularity, wherein the quarter, half, and completed year is absolutely dependable, and wherein seasonal planning and calculation of products will save labor, money and time—will have great advantages.

## Government

The task of the government will be lightened when our Federal Government begins its fiscal year always on the same day and date. A perpetual calendar will materially simplify comparisons of statistics. The records on customs receipts, income taxes, internal revenue collections and interest paid or received will be more easily computed. And for the various government departments which require accurate quarterly financial statements, the advantages of The World Calendar, with its equalized quarterly division of even 3 months, 13 weeks or 91 days, are apparent.

## Statistics

For statisticians and accountants, a calendar that is orderly and steady, one which is based on a set mathematical formula, will make comparability possible. This will greatly lessen midnight oil and the mental and physical strain due to our present calendar difficulties. Statistical reports will be more easily compiled and be more accurate because the monthly and quarterly patterns are not changed. Where the temporary auxiliary system of 13-week periods has been adopted, The World Calendar stabilizes the 13-week periods in each quarter-year, retaining the convenient 3 months.

## Education

In the educational world, the special benefit to the school year is the stabilization of Labor Day Monday, always on September 4. The regular opening of schools on the same weekday and month-date each year will lead logically to the orderly closing of the first and second terms on their regular weekdays and month-dates. The same reasoning applies to countries where there are three school terms in the year. With Labor Day always placed on the first Monday, September 4, there will be no need either to cram or to lag in the studies between the opening day of school and Christmas. This, among other advantages, will be immediately recognized by both faculty and students. The more uniform vacation periods possible under this new time-schedule will make it easier to unite families which, in the past, were too often separated during vacations because of the present haphazard calendar. Much of the unnecessary time expended in arranging study schedules, annual programs, vacations, and sports events will be made available for more worth-while activities.

## Law

For lawyers, the benefits The World Calendar will bring are incalculable. Their obtuse and wordy language will in great measure disappear. The "first Tuesday after the first Monday in November" will become the simple, straight-forward November 7. The first Monday in October for the opening of the Supreme Court of the United States becomes October 2.

The year, meaning 12 months; the half-year, meaning 6 months; and the quarter-year, meaning 3 months, will not be restricted to months only. Their number of weeks and days will henceforth be included in their terminology. Thus every year means 12 months, 52 weeks, 364 days; half-years 6 months, 26 weeks, 182 days; and quarter-years 3 months, 13 weeks, 91 days. The 365th day, essential to the calendar, the extra Saturday World Holiday, December W, and the 366th day in leap years, June W, will be legally considered as World Holidays throughout the globe. And who can measure the advantages these will bring?

## Finance

Equally important to finance is the good result for the banker, the depositor and the creditor. With every quarter-year equal, having exactly 91 days, a quarterly note will be an exact one-fourth of the annual loan. A 30-day note will be one-third of a quarter year, because in the 31-day month in each quarter-year the extra day is always a fifth Sunday and thus no business day. The advantages to the borrower and lender are obvious with the usage of The World Calendar of equal quarters.

## Radio

The radio business is unique in that its contracts are based on a 13-week quarter-year period. This is meticulously maintained in The World Calendar with the added benefit that the quarter-year periods always begin on the same day and date. And the new World Holidays take care of that extra 365th and the quadrennial 366th day which in the past have been a constant source of annoyance.

## Transportation

Complicated planning of time-schedules for the various means of transportation, whether by road, air, sea or rail, will be simplified when holidays, vacations and other outstanding events are known to fall on the same day and date year after year. Business dates can be arranged so that these do not interfere with college and university reunions, for it will be known that a Yale, Harvard, Notre Dame or Leland Stanford commencement will fall on a definite day and date. The standard World Calendar, like the standard clock, will bring untold advantages into family, social, business and national life not realized at present.

## Summer

Summer seasons beginning with Memorial Day, the new May 27th, and closing with Labor Day, September 4 (a period of fourteen weeks) will enable hotel keepers and concessionaires, cottage renters and summer clubs to plan with greater assurance.

## Entertainments

Social entertainments and operas, concerts, theaters and movies; art, flower exhibits and horseshows, boxing and ball games, and other similar events will be planned more easily with overlapping dates reduced to a minimum. Everything will move more simply and smoothly.

## History

Historians who deal with the records of the past need not be apprehensive that a confusion of dates will result from a "change of style." Unlike the Gregorian reform, which altered the calendar by dropping to 10 days (11-13 according to the time of its adoption by nations), only one or two dates within each of the six months, between February 28 and September 1, in the new time-plan will be affected by the conversion. The days in the other six months, between September 1 and February 28, remain exactly as they are.

An important feature in the perpetual calendar is of special interest to historians:

"Future historians would thank us if we could place the dates so that the weekday and the month-day always came together. For example, if an historian finds the date September 13, he immediately knows that it was on Wednesday in the middle of the week. This might give a different interpretation to an event than would be given if the date were September 11, which he would know happened directly after a Sunday." [32]

As the calendar is now, it is quite impossible for future generations to remember that Hitler began World War II, by his march into Poland, on *Friday* September 1, 1939; that England declared war on *Sunday* September 3, 1939; and that Japan attacked the United States in the Hawaiian Islands on *Sunday* December 7, 1941. We shall constantly have to refer to a so-called "perpetual" calendar to find the right day of the week with the date. Or we shall have to resort to historical reference books that are not readily at hand and in which, quite often, weekdays are not mentioned in connection with month-dates. It is too little realized and understood that the *day* as a designating time-unit exerts its own peculiar influence and thus must not be lost, as it is now. The future may reveal many things with the stable and regular calendar. When we consider the convenience of standard time in our radio programs, what potentialities may not the stabilized and well-planned World Calendar disclose?

## Birthdays

How will the people be affected whose birthdays fall on the three lost days—March 31, May 31 and August 31? Where lies the advantage, if they lose their birthdays? Their birthdays will not be lost, as they may be observed on the day before, just as leap-year children observe February 29 in leap years, and on the 28th in ordinary years.

[32] Fox, Dr. Dixon Ryan, *Journal of Calendar Reform*, Volume 10, p. 46.

There exists in England an historical document, 1236 A.D., wherein King Henry III decreed at Westminster a law which reads:

"The day increasing in the Leap Year shall be taken and reckoned in the same month wherein it groweth and that day, and the day next going before, shall be accounted for one day." [33]

In this quaint language, King Henry planned that anyone born on a leap day should have his regular birthday on the day previous in all but leap years. This human provision serves as well today as when it was first written, seven centuries ago. Individuals who were born on the 31st day of March, May and August in the present calendar are invited to observe their anniversaries on the day before, the 30th. *Thus their birth-months are not changed.* And those born on December 31 will observe their birthdays always on the new World Holiday, Saturday, December W, the Year-End Day.

## Calendar Makers

Objections have been raised to a perpetual calendar on the ground that it will curtail the market of calendar manufacturers who sell calendars every year and so discriminate against their particular business. It has been pointed out, however, that there is nothing quite so useless as a last year's calendar and an overstock of calendars at the close of the year is a loss to manufacturers, retailers and distributors. This loss is wiped out by the perpetual calendar. Perpetual calendars offer artistic opportunities in design and material. They can be fashioned in gold or silver, bronze or marble,

[33] *Journal of Calendar Reform*, Volume 12, p. 135.

with precious or semi-precious stones. They can be made elaborate or simple, stationary on the wall, or portable for desk or table use, wallet or purse.

Clock industry does not suffer nor complain because clocks have the same dial. Clocks are permanent and because this is so they are beautiful and durable. Notwithstanding, clocks will wear out and be in steady demand; likewise the calendar, although perpetual in arrangement, requires annual calendar pads, diaries and notebooks. And obviously calendars and almanacs containing information on the rising and setting sun, the rise and fall of tides, and the phases of the moon, will change every year even though the calendar itself remains the same. The steadiness of the calendar will simplify everything.

## Clock Makers

And here a word to the clock makers. It is believed that a 24-hour clock is inevitable. To count the day by 24 hours is more reasonable and lends itself to fewer mistakes than does our present clock time of 12 hours A.M. and 12 hours P.M. That the United States Government recognizes this is evident in the military service where everything is based on the 24-hour system. Many men in the service will undoubtedly bring home the 24-hour clock idea making its use customary. Does not the willingness of the government to change clock time for something better indicate a like willingness to change calendar time for something better?

## Order with Variety

It has also been said that a perpetual calendar is regimented and people dislike regimentation and monotony. Well, so does the calendar, which, like people, enjoys order with variety. We need only to observe the regular recurrences of the different seasons, the variety of fruits and vegetables, trees and flowers, that have their definite time of growing and blossoming. Nature abhors rigid monotony but loves ordered variety. The equal quarters have in their arrangement a similar order with variety.

### On the Square

These and many other advantages are the result of a calendar built on-the-square. The principle of equal quarters, wherein exists perfect co-ordination, co-operation and justice, makes it easier for everybody and everything to be on-the-square. It will bring nearer the promise of the spiritual city, new Jerusalem, that lieth foursquare, where within its gates, nations and peoples live in glory and honor.[34] The world needs the square deal to become a reality, as exemplified in The World Calendar, a plan ready and available, eager to serve for the common good of all mankind.

[34] *Revelation* 21:10-26.

# 16

# HOLIDAYS

T HE AXIOM "All work and no play makes Jack a dull boy" applies to the calendar as well as to the individual. The ordered rhythmic breath of Earth expresses itself best in a smoothly flowing and tranquil calendar where there are occasional days of relaxation and joy, praise and prayer, distributed through longer series of work-days. Holidays keep the calendar human. Is there anyone in the world who would wish to forego them? No, for holidays are essential.

There is not a nation in the world which does not observe holidays, irrespective of whether these are religious or national. In most countries the greater number seem to be religious. In the United States they are chiefly national in character, commemorating national events and honoring national heroes. But here, too, three significant religious days are observed annually—Easter, Thanksgiving and Christmas.

In a concise and well-planned booklet annually compiled by the Guaranty Trust Company of New York, the bank and public holidays of the various nations are carefully listed. One is amazed to find in the 1942 issue that there

# HOLIDAYS

## Under The World Calendar

---

### New Year's Day

Sunday, January 1

### Washington's Birthday

Saturday, February 11 (the original date of his birth) or Wednesday, February 22

### Lincoln's Birthday

Sunday, February 12, to be observed Monday, February 13

### Easter Sunday

April 8, subject to church decision

### Memorial Day

Thursday, May 30, or Monday, May 27

### Leap-Year Day

extra Saturday, June W

### Independence Day

Wednesday, July 4, or Monday, July 2

### Labor Day

Monday, September 4, or Monday, September 11

### Columbus Day

Thursday, October 12, or Monday, October 9

### Election Day

Tuesday, November 7, or Monday, November 6

### Armistice Day

Saturday, November 11

### Thanksgiving Day

fourth Thursday, November 23, or fourth Monday, November 27

### Christmas

Monday, December 25

### Year-End Day

extra Saturday, December W

are 238 out of our 365-day year that are observed as holidays somewhere on this globe. Fortunately, these 238 holidays are not observed everywhere. What a disturbing thought were we to have a preponderance of play-days rather than work-days!

According to this compilation, the Union of Soviet Socialist Republics has the fewest holidays, four, England has seven, the United States, in comparison, has eleven, whereas India observes 31 holidays somewhere in her vast country. And Thailand, formerly Siam, the latest nation to have adopted the Gregorian calendar in 1941, is unique in her celebration of holidays in that she includes quarter-year and half-year holidays.

Because the writer is more familiar with the holidays of the United States of America, she will permit herself to discuss these.

## New Year's Day

As the calendar always begins with January 1, this day will receive first consideration. There is an underlying significance to the New Year's Day in The World Calendar as it will always come on *Sunday,* January 1, on the first day of the week, the day of the Lord. This auspicious beginning cannot but exert a valuable and far-reaching influence upon mankind. It is as if we, the people, will always start the new year right. Thus we begin every year with worship and song in religious and communal gatherings. The following day, *Monday,* January 2, is always *the first working day* in each new year. We shall see toward the end of the chapter why the day immediately following the Sunday is *not a holiday* according to previous custom.

## February

We now progress to February which in The World Calendar will have 30 days, attaining its rightful share, and as such receiving the same consideration given the other 30-day months. No longer is February unjustly deprived of its full quota of days. This month has the signal honor of marking the births of two great patriots: Washington, the father of his country, and Lincoln, the preserver of the Union.

February 11, the true birthday of Washington, as we have noted in the chapter on the Gregorian calendar, will be annually observed on Saturday, with the following day Sunday, February 12, Lincoln's birthday, celebrated as is customary on the following day Monday. Thus the two great American patriots may more closely unite Americans through the medium of their joint week-end birthdays, fostering thereby a greater spirit of unity, comradeship and service.

## March and April

March, our blustering month, is devoid of holidays and April, too, will not enjoy a holiday, although within that month comes the joyous Easter Sunday, which it is confidently expected will be observed on the second Sunday in April, April 8.[35] This fixed Easter date will remove all the previous uncertainties as to the Sunday for its observance and the adherence to the inconstant moon will have vanished. The regular and tranquil Easter date will be a blessed boon to schools and universities, shops and business, family and social life, parishioners and churches.

[35] See chapter 17, p. 116.

## May

In the month of May comes Memorial or Decoration Day, which was established after the Civil War, or the War between the States, through the efforts of General Logan, a Union Army officer. He felt it proper for the United States to express its gratitude and to commemorate its heroes. The first Memorial Day was observed May 30, 1868, a Saturday. Because of the friction then existing between the North and the South it has not enjoyed the national observance it deserves. With the World War I and the present World War II it is ardently hoped that Memorial Day will truly come into its own, on which holiday the entire nation can unitedly pay tribute to its noble dead. This day, observed in the North on May 30 and in some Southern states on different days, may fully enjoy its new date, May 27, the last Monday in May. It would make another lengthened week-end holiday. With a fixed Easter, April 8, Whitsunday, May 26 and Whitmonday, May 27 (a holiday in other countries), this new week-end holiday could be observed throughout the world, regardless of the different meaning.

## June

This is the month of the Leap-Year Day so that it will have a holiday every four years. It will be observed on the extra Saturday following June 30, considered a World Holiday, with its date June W. By placing this day always at the close of the first half-year, both of the half-years in leap years will have the same number of days—183—and perfect balance is secured. On this unique date a wonderful opportunity is offered nations and peoples to unite in joint observance in a spirit of peace, good will, recreation, amity, co-operation, a giving and a sharing.

### July

America celebrates her birthday on July the Fourth, a Wednesday, in the new civil calendar. As a Wednesday holiday breaks awkwardly into the week it is at all times unwelcome. Why, then, should America's proud and glorious birthday be so poorly treated? There is historical authority that the Declaration of Independence was submitted to the Federal Congress on July 2 when it received twelve yeas and no nays and was approved on July 4. America's birthday, coming as it does in our hot summer days, will offer more wholesome enjoyment and recreation to every American when celebrated in the new civil calendar on Monday, July 2, rather than sandwiched between two business days.

Thomas Jefferson was the chairman of a committee of five selected to draft the Declaration of Independence and among the members was John Adams. The latter wrote to his wife on July 3, 1776, the day after the Declaration was submitted to Congress: "I am apt to believe that it will be celebrated by succeeding generations as the great anniversary festival." Strange as it may seem, this prognostication may yet come true. Certainly we are living in the midst of changes even stranger than this. It is of marked interest that these two men, signers of the Declaration, who served their countries as Presidents, and were political rivals yet staunch and loyal Americans, passed this earthly scene on Thursday, July 4, in the same year 1826.

*August*

This vacation month has no holiday. It is the recreational month in which Americans prepare for the coming activities of autumn and winter. Of course it is realized that in countries below the equatorial belt, the summer months of the North become the winter months of the South.

*September*

This month dedicates its first Monday to Labor, which in the new civil calendar comes on September 4. It signals the end of summer vacation days. School children return to trot back and forth to their lessons preparing themselves to become worthy citizens. Summer hotels prepare to close and the annual exodus from country to city life begins.

*October*

One holiday, that of Columbus Day, October 12, commemorative of the discovery of the Western Hemisphere, is celebrated in this autumn month. It is worthy of greater recognition than it receives. But here, too, it falls on a weekday, Thursday, and because of this, interferes too much with schools and business and therefore could well be observed on the previous Monday, October 9. It is the autumn month of the North, the spring month of the South Americas, the in-between season admirably suited for hemispheric celebration. It would be an annual festival which all the Americas, North, Central and South, could observe as America's friendship day, cementing more closely inter-American relationship. Especially so, as the name Christopher Columbus is suggestive of the Christ and the dove (colomba). Does not this invite contemplation?

*November*

There are three national holidays observed in this month—Election, Armistice and Thanksgiving.

To observe Election Day in The World Calendar on the first Monday, November 6, as Labor Day is observed on the first Monday in September, would seem to be a great advantage over the present custom which observes it on the "first Tuesday after the first Monday in November." To end the hectic, intense political campaigns on Saturday night, thereby affording the citizens a Sunday for quiet study, analysis and contemplation for voting the following day, would, in the opinion of many, be of incalculable value to the entire country.

Five days later on Saturday, November 11, Armistice Day is observed. To have it come naturally on the half-working day, Saturday, minimizes its importance and does not seriously interfere with our many other activities. The quiet one-minute silence on Armistice Day, as observed by the English, is more fitting. Let us hope that at the end of the World War II, Armistice Day will give way to the more momentous day on which peace will be signed.

The fourth Thursday of November is our American Thanksgiving. The first Thanksgiving was proclaimed by Governor William Bradford of Plymouth in 1621, in gratitude that the first harvest had been so bountiful, "had all things in plenty." Through varied and irregular observances Thanksgiving Day received its first observance as a national holiday by a presidential proclamation of George Washington, who designated it for Thursday, November 26, 1789. This day was set aside:

"in rendering unto Him our sincere and humble thanks for His kind care and protection of the people of this country ... to render our National Government a blessing to all the people by constantly being a government of wise, just and constitutional laws, discreetly and faithfully executed and obeyed."

Again there were frequent lapses in observing this holiday until Abraham Lincoln, the preserver of the Union, in the dark days of the Civil War, proclaimed in 1863 that Thanksgiving be annually observed on the *last* Thursday in November. And, oddly enough, it was again a November 26.

The last Thursday, selected by President Lincoln, established the holiday for Americans and thus it was observed for about three quarters of a century. Then in 1941, by Act of Congress and approved by another war President, Franklin D. Roosevelt (shortly after the tragedy of Pearl Harbor), Thanksgiving was definitely set on the *fourth* Thursday in each November, so that in 1942 its first observance on the fourth Thursday was once more November 26.

These three November 26ths marked notable steps in the history of this uniquely American day. And yet this remarkable triad of dates does not find its prototype in Thanksgiving itself. It still wanders as to its date, notwithstanding that the day and the month are settled. With the adoption of the perpetual World Calendar, Thanksgiving would receive its logical date, November 23, and the triad be complete. It would always be observed on the *fourth Thursday,* its day; on the *23d,* its date; and in *November,* its month.

It is hoped that the President of the United States, Franklin D. Roosevelt, during his present Administration will complete his initial action and give to our country and, through the United States of America, to the world the new perpetual World Calendar. It would be an act that would resound through the ages and be a blessing to civilization.

## December

Christmas is the inspiring December holiday and naturally falls on Monday, December 25, giving a most welcome week-end. Saturday would end the loving gift-buying; Sunday would offer spiritual preparation; and Monday would be the day of its happy celebration.

With the approval and adoption of The World Calendar two new World Holidays appear—the Year-End Day, that extra Saturday, December W, and the Leap-Year Day in leap years, another extra Saturday, June W. *Because the Year-End Day, a holiday, precedes Sunday, January 1 (New Year's Day), it naturally voids the usual custom of celebrating a Sunday holiday on a Monday.* Monday, January 2, would obviously be the first working day of the year.

# 17

# EASTER

THE NEW CIVIL CALENDAR is now ready to be launched into the world, full of hope, promise and aspiration. Our story is not fully complete, however, without the mention of a fixed Easter. And yet on this subject I hesitate, realizing that this feast day belongs to the religious life of man and the decision to stabilize its date rests with religious authorities. As a layman, one who has had remarkable privileges in talking with religious leaders of many different church groups, I may be permitted to express a few of my observations.

It is obvious, of course, that the steady and ordered World Calendar can function without a fixed Easter since it is a civil calendar, and religious feast days may still be movable or fixed as the various churches decide. To harmonize our civil with the religious life, however, would greatly enrich and inspire man, and the civil World Calendar, too, would be imbued with a greater tranquillity were Easter to have its constant date. It is not possible to have a fixed Easter date within our present changeable and irregular calendar. And even were Easter to be stabilized in our present calendar it would still be a wandering feast day, although restricted to a cycle of 7 instead of 35 days.

Just as from a far-gone past came the months of irregular days with their accompanying changing dates in the civil calendar, so there was bequeathed to the churches the changeable Easter in the religious calendar.

The early Christians found themselves in a quandary regarding Easter. They were deeply concerned in retaining the first day of the week, Sunday, the day of the Resurrection, for its annual observance, yet equally anxious to associate it with the Passover moon. As of old, tradition and custom were powerful influences. A compromise was effected at the Council of Nicea in 325 A.D. establishing Easter on the first day of the week, Sunday, but the Sunday was to be adjusted according to the spring full moon.

Easter was to come on the first Sunday after the first full moon of spring which falls on or after the spring equinox, March 21; and should the full moon come on a Sunday then Easter is to be celebrated the following Sunday. The ecclesiastical full moon, however, is *not* the astronomical full moon as we observe it, but a full moon that follows 14 days after the new moon. Easter is celebrated, therefore, on the first Sunday after this ecclesiastical full moon on or after March 21. The full moon was selected in ancient days because pilgrims needed the moonlight for their safety in travel.

This year, 1943, Easter was observed on April 25, the latest date on which it can fall. This extraordinary event happens but once in a century. Heretofore it has occurred in 1546, 1641, 1736, 1886, and, if no improvement is made, then again in 2038. What a complication of dates confronts the church calendars!

The early Christians had hoped that by their method, a coincidental date between the Jewish Passover moon and the

Christian Easter might be avoided. And in the main they were successful. It is interesting to observe that the Latin countries call Easter Pascua, Pasque or Pâques, derivatives of Passover, whereas the Anglo-Saxons call it Easter or Oster after the spring goddess, Eostre.

Within the years 1940 and 1946, Easter is particularly vacillating, demonstrating by its wanderings the need for improvement. In the table below we see that in 1940 Easter fell on March 24; in 1941 on April 13; in 1942 on April 5; in 1943 on April 25. Then it is thrust back to its middle-of-the-way date, April 9, in 1944; only to renew its harried wanderings to April 1 in 1945, followed a year later by April 21.

THE WANDERING EASTER

(Easter Dates from 1930-1946)

| | |
|---|---|
| 1930—April 20 | 1938—April 17 |
| 1931—April 5 | 1939—April 9 |
| 1932—March 27 | 1940—March 24 |
| 1933—April 16 | 1941—April 13 |
| 1934—April 1 | 1942—April 5 |
| 1935—April 21 | 1943—April 25 |
| 1936—April 12 | 1944—April 9 |
| 1937—March 28 | 1945—April 1 |

1946—April 21

Great Britain's activity for a fixed Easter, which Parliament set for "the first Sunday after the second Saturday in April," (provided that other Chrisian Churches agreed) paved the way for similar action to be taken by the League of Nations. Among the many Christian nations approving League action was the United States of America, on "the opinion that the common good, from a purely economic and social standpoint, would be served by such stabilization."

To the writer this activity on a religious matter was unwise and the choice of date unfortunate. The date endorsed by the League of Nations would always be Sunday, April 15, with Good Friday observed on April 13. With the Crucifixion regularly observed on the 13th it would seriously tend to accentuate and strengthen the superstitious influences—a lamentable condition which the churches would hardly wish to countenance. Then, too, Easter always coming on the 15th of April would interfere with mid-month financial settlements, introducing an undesirable and material note. These two dates are not the very best that could be selected. According to astronomers and historians the general opinion seems to be that the Resurrection occurred on the first day of the week, April 9, 30 A.D., on the 99th day of the year. In The World Calendar, Sunday April 8, *the second Sunday in April,* would thus stabilize Easter nearest the approximate historical date and bring Good Friday on April 6. Then there is another religious advantage for these dates. There is no other major feast or fast day that would interfere with them. Sunday, April 8, would also be the 99th day of the year.

With Easter observed on the suggested April 8, Ash Wednesday would come on February 22. As George Washington's birthday would be celebrated in The World Calendar on the original date, February 11, Ash Wednesday would not interfere with it. Passion Sunday would come March 24, followed immediately by "Our Lady's Day," Monday the 25th. Palm Sunday would fall a week later on April 1, Good Friday, April 6, culminating with Easter, April 8.

Progressing onward, Ascension would be Thursday, May 16, with Whitsunday or Pentecost, May 26.

Furthermore, Trinity Sunday would be the first Sunday in June, the 3d, with the last of the Sundays following Trinity coming on November 26. Advent would then be the first Sunday in December, the 3d, foretelling the coming nativity on the four Sundays in December. The last Advent Sunday would be Christmas Eve, heralding Christmas Day.

Would not this fixed arrangement of feast and fast days on their regular days and dates *simplify* the church calendars?

Seldom has there been offered to Christianity so timely an opportunity to demonstrate that a world-wide religious agreement is possible. Agreement on a fixed Easter would be an inspiring event indeed, uniting Christendom as one.

A good beginning has been made for general agreement by the Protestant Episcopal Church, the Methodist Council of Bishops, the American Lutheran Church, the Universal Christian Council for Life and Work, all approving a constant Easter. And the Western European representative of the Eastern Orthodox Church is also in favor of it. A statement emanating from Pope Pius X in 1912 is particularly significant. "The Holy See declared that it made no objection but invited the civil powers to enter into an accord on the reform of the civil calendar, after which it would willingly grant its collaboration in so far as the matter affected religious feasts." [36]

Were the Vatican, center of the Roman Catholic Churches;

[36] Massiani, M., "Question of Easter and Calendar" in *Catholic News,* March 27, 1937; *Journal of Calendar Reform,* Volume 17, p. 85.

Lambeth Palace, fountain-head of the Anglican faith; Constantinople, seat of the Eastern Orthodox Church; Geneva, headquarters of the Universal Christian Council for Life and Work (the provisional World Council of Churches); and New York City, home of the Federal Council of Churches of Christ in America—were all these Christian groups to call concurrently a religious conference for the stabilization of Easter, it would truly inspire the world. It would be as if "a new heaven and a new earth" had come and would bring *new* courage, hope, and faith to mankind. It would be a glorious demonstration that a united Christianity had become a shining beacon for all. As the planets revolve around the sun and show fealty to it; as the time-units revolve within the calendar and show fealty to the solar seasonal year; so Christendom revolves around Jesus, the Christ, and shows fealty to God.

But where is such a conference to meet? When one realizes that Easter was established in the Eastern Hemisphere, it would seem eminently right that Christendom might consider assembling in the new world—the Western Hemisphere—to establish a fixed Easter date. And the United States of America, wherein all the churches have *equal rights* to worship God in their own way, *unhampered and free,* and wherein no one church possesses greater privileges than another, would seem to be the ideal country for such a Christian conference, and Philadelphia,[37] the city founded by William Penn in the spirit of brotherly love, the fitting place.

Christianity would realize as never before the Psalmist's

---

[37] Philadelphia is namesake to that city of churches in *Revelation* to which words of praise were given; *Revelation* 3:7-12.

song: "Behold, how good and how pleasant it is for brethren to dwell together in unity!" [38] and it would be a fulfillment of the words of Jesus, "For where two or three are gathered together in my name, there am I in the midst of them." [39]

I like to think of religious feast days as radiant stars in the rhythmic succession of our calendar days, annual reminders of the intimate relationship that exists between God and man. Religion knows no frontier and like the Creator is limitless in its wide horizon.

For Christianity to act on a fixed Easter will bring peace on earth and good will toward men, nearer to fulfillment. Future generations will look with pride and gratitude to this generation and decade, for having had the wisdom and courage, notwithstanding storms of war, to rise above these turbulent conditions and lay a milestone on the forward march of civilization's progress. One that marked a greater freedom and emancipation in the life of man, leading him ever onward to greater unity and understanding fellowship.

[38] *Psalm* 133:1.
[39] *St. Matthew* 18:20.

# 18

# ADOPTION

Seldom, if ever, is such a rare and unusual opportunity offered to any nation, people, ruler or leader as that of revising the civil calendar. We have noted from the past pages of this book that in a long period of six thousand years only five such instances are recorded in the sun calendar and that, alas, in our more recent day a group of international representatives, by waiving it, lost it.

Fortunately for the progress of civilization and the welfare of man, another opportunity is knocking for entrance, and this time before the new world in the Western Hemisphere. We cannot conceive, nor do we believe, that the Americas would wish to follow the weak and hesitant footsteps of the receding League of Nations which allowed its opportunity to pass. For has not the Western Hemisphere come to its own rightful heritage to accomplish for Time, what the ancient Egyptians and the Southern Europeans of the Eastern Hemisphere did, in their day and era? And among the Americas there appears today none more fitting to take the leadership and to act for The World Calendar than the United States of America. Comprised as she is of peoples from all parts of the globe, she is well equipped in thought, mind and spirit to initiate it.

To adopt the harmonious and equitable World Calendar without the disturbances caused by the Julian calendar (when a year of confusion of 445 days was a necessary transition from the moon to the sun calendar) and the present Gregorian calendar (that cancelled ten days to adjust the calendar with the seasons), it is best to choose a day, date, month and year when both the old retiring and the new incoming calendars glide smoothly together. Such an opportune time was January 1, 1933 and again January 1, 1939; both were New Year days.

The next easy progression will be Saturday, December 30, 1944, when both the old and the new calendars meet. The following day, Sunday, December 31 in the old Gregorian, will become the new World Holiday, extra Saturday, December W, in the new. The World Calendar will then proceed on its course Sunday, January 1, 1945. In this way the calendar keeps its scientific accuracy with the seasons, for which in the past it had labored so long.

Other dates on which the old and the new calendars meet, according to Henry W. Bearce, Chief of the Division of Weights and Measures, National Bureau of Standards, United States Department of Commerce, are Sunday, July 1, 1945 and Friday, March 1, and Wednesday, May 1, 1946. In the words of Mr. Bearce, "The World Calendar could be put into effect without any perceptible jar to our social or economic structure on days and dates at the beginning of a month agreeing with the Gregorian calendar. This occurs in 1945 and 1946. The change should be made *before* 1947. The following years—1947, 1948, 1949—are the lean years in which no such coincidental date occurs. The next date would be Sunday, January 1, 1950."

# HOW EASY THE ADOPTION
# ON THESE DATES

## 1944

| *Gregorian Calendar* | *The World Calendar* |
| --- | --- |
| Saturday, December 30 | Saturday, December 30 |
| Sunday, December 31 | Extra Saturday, December W |
| | (a World Holiday) |

## 1945

### *The World Calendar*

Sunday, January 1

OR

| *Gregorian Calendar* | *The World Calendar* |
| --- | --- |
| Sunday, July 1 | Sunday, July 1 |

## 1946

### (not as convenient, change comes in the week)

| *Gregorian Calendar* | *The World Calendar* |
| --- | --- |
| Friday, March 1 | Friday, March 1 |
| Wednesday, May 1 | Wednesday, May 1 |

## 1947, 1948, 1949

### (no coincidental dates, the lean years)

## 1950

### (an unjustifiable delay)

| *Gregorian Calendar* | *The World Calendar* |
| --- | --- |
| Sunday, January 1 | Sunday, January 1 |

In Mr. Bearce's opinion the best date for adoption would be Sunday, July 1, 1945, the beginning of the third quarter and second half of the year. On this day, date, month and year, The World Calendar would glide smoothly and easily into our daily lives and affairs with no disturbance whatsoever. Peoples all over the world could retire on Saturday, June 30, 1945 in the old calendar and awaken the next day, Sunday, July 1, 1945 in the new without any dropping or adding of days. This calendar change would be appreciably easier than the change from Standard Clock Time to Daylight Saving, War or Summer Time when the clocks are changed by one or two hours.

The first notable change in the new calendar would come nine weeks later when there is no longer a 31st day of August. Birthdays and other anniversaries coming on the old date could be observed the day before, August 30. Then would follow September, October and November with no changes. The last day of December, the 31st, would become the new Year-End Day, the extra Saturday World Holiday, December W, with the following day automatically observed as Sunday, January 1, 1946 in The World Calendar.

It has been suggested that where advanced publications, such as almanacs and ephemerides, have been printed according to the Gregorian calendar, supplementary sheets with The World Calendar data may be added to indicate the changes that occur between February 28 and September 1.

There is an opinion that the new civil calendar must be universally adopted and yet calendar history does not substantiate this. No reform, no good thing has ever been universally accepted. Nations and peoples have not acted that way in the past. The one movement perhaps that has come

closest to universality was Standard Time adopted in the
United States in 1883, but there, too, certain peoples opposed
it on the ground that it was interfering with "God's Time."
And the recording of clock time, whether on the 24-hour
Continental system or the 12 A.M. and 12 P.M. method of
other countries, is still an unsettled question.

In discussing Standard Time it is decidedly interesting
that it was introduced and put into operation by the American Railway Association and used by the citizens without
need of legislative action. The United States Congress recognized it 35 years later, as late as 1918, when it directed
the Interstate Commerce Commission to establish the various
time zones in the country.

Weights and measures, currencies, trade treaties are not
universally adopted. Why, then, should the universal acceptance of the new civil calendar be the criterion of its adoption? In the words of Samuel Johnson:

"Nothing will ever be attempted if all possible objections must
be first overcome."

Let us be fair and just, and allow no wishful thinking to
delude us into delaying acceptance because of this erroneous
theory.

Of course it would be highly desirable to obtain universal
adoption at once. As the United States of America, home of
democracy, believes in world neighborliness and human
brotherhood, she is in an unusually strong position to take
the lead and with other American nations in the Western
Hemisphere sign a treaty accepting The World Calendar,
1945. The approval and adoption by the Western Hemisphere would bring general world acceptance nearer and

the new civil calendar be "one and the same for all the world."

Today when every second, minute, hour and day are so vital we must not delay. The future is far more unpredictable today than it has ever been in the past. Meticulous timing, well-laid plans, and constructive thinking are essentials, regardless whether the world is at war or at peace. Time, the calendar, Earth's breathing, goes on notwithstanding. But we are and should be intensely concerned that Earth's breathing is ordered and rhythmic, so that it helps in every way. A better calendar adopted now *in the present* is a constructive and wise procedure upon which to prepare and build a better future in every field of endeavor.

The hope of the future of civilization depends upon *our acts today*. And among the acts of today belongs undeniably the adoption of the new and better civil calendar—The World Calendar. For nothing is built, nothing undertaken, nothing achieved, without the element of Time—Earth's breathing. Without breath we cannot live and without Time there is no motion. Breath and motion are life givers, and our life source and sustenance must be of the fullest and best, whereby we may continuously develop. And the inseparable companion of Time is the calendar.

But how can action be taken?

*First:* The President could appoint a small committee to study calendar revision; its findings being favorable, it could then be presented with the President's approval to the Department of State. Such action would be the basis for an inter-American conference to obtain official approval for The World Calendar.

*Second:* The President and the Department of State could

invite the six Latin-American countries—Brazil, Chile, Mexico, Peru, Panama, Uruguay—which have already officially approved it, together with the other Americas, to join in an inter-American conference, or, through the Pan American Union, to deal with the subject toward its adoption in 1945. Strong leadership emanating from Washington would undoubtedly point the way for other countries and peoples to follow.

Should the delegates of the various American countries be prevented by war conditions from coming to Washington for this conference, then the various Latin-American Ambassadors in Washington could be given authority by their respective countries to act in their behalf for the approval and adoption of the new perpetual civil calendar of 12 months and equal quarters. Ratification of The World Calendar could be accomplished by a treaty signed by the nations of North, Central and South Americas, and in the United States approved by the Senate.

*Third:* The countries in Europe and Asia that have or have not as yet endorsed the calendar could likewise request their Embassies to act and join the Americas in approving the new time-plan, and likewise confirm their position by the signing of the treaty.

We can rest assured that the Americas, like the countries of old in Africa and Europe, will not fail but will courageously take the initiative in giving the world this superior time-system. They will be modern trail blazers sending the calendar ever onward on its forward march to assist the many other new and better systems that are in the making.

Are these not the very aims of peace and of maintaining peace? When we say peace I like to think of the Chinese word for peace, *Ho Ping,* meaning harmony-equity. And certainly harmony and equity dwell within The World Calendar.

There is every reason to believe that the orderly and beautifully symmetrical World Calendar "the very best, with all conditions considered" will truly be a friendly Federation of Time and as such aid the world in ultimately achieving a friendly Federation of Nations, in the economic field a friendly Federation of Economics (trade, currency and finance), and in the social field a friendly Federation of a World Society.

I close with a deep feeling of gratitude to Divine Providence for the privilege of working for a cause that will contribute toward making a better world, making life more beautiful, and making my fellowmen happier.

# CHRONOLOGY

## (This list comprises only the salient facts)

| | |
|---|---|
| B.C. 4241 or 4236 | Origin of Egyptian Sun Calendar |
| B.C. 3251 | Date contended by Jotham Johnson |
| B.C. 753 | Founding of Rome (A.U.C. 1) |
| B.C. 715-672 | Numa Pompilius gives Roman Calendar 12 months and establishes the College of Pontiffs |
| B.C. 304 | Cneius Flavius liberates calendar |
| B.C. 238 | Ptolemy Euergetes proposes leap year day; rejected |
| B.C. 153 | New Year is changed to January 1 |
| B.C. 63 | Julius Caesar is elected Pontifex Maximus |
| B.C. 46 | Year of confusion (445 days) |
| B.C. 45 | Julian Calendar introduced |
| B.C. 8 | Augustus Caesar adjusts leap year rule, certain leap years eliminated |
| A.D. 1 | Christian era begins |
| A.D. 8 | Leap years reinstated |
| A.D. 238 | Censorinus deals with measurement of time |

A.D.  321        Constantine introduces the week into the civil
                 calendar

A.D.  325        Council of Nicea

A.D.  325-385    Macrobius' account of calendar in *Saturnalia*

A.D.  532        Dionysius Exiguus establishes Christ's birth as
                 1 A.D.

A.D.  1582       Gregorian Calendar introduced

A.D.  1731/32    George Washington's birth

A.D.  1752       Great Britain adopts Gregorian Calendar

A.D.  1834       Abbé Mastrofini proposes perpetual calendar
                 with extra stabilizing days

A.D.  1849       August Comte devises 13-month calendar

A.D.  1887       Astronomical Society of France awards prize
                 to perpetual 12-month calendar of equal quar-
                 ters; rejects 13-month calendars

A.D.  1900       Non-Roman churches at Eisenach study calen-
                 dar revision

A.D.  1910       Congress of International Chamber of Com-
                 merce adopts resolution in favor of calendar
                 revision

A.D.  1912-1914  Calendar revision continued

A.D.  1914       International Chamber of Commerce requests
                 Swiss Government to study question and to
                 call a conference. World War I prevents fur-
                 ther activities

| | |
|---|---|
| A.D. 1922 | International Astronomical Union's Commission 32 recommends perpetual 12-month equal-quarter calendar |
| A.D. 1923 | League of Nations includes study on calendar revision and a Special Committee is formed |
| A.D. 1924 | Anglican, Eastern Orthodox, Roman Catholic religions state there are no dogmatic objections to calendar revision and fixed Easter |
| A.D. 1925 | International Chamber of Commerce approves League activity for calendar revision |
| A.D. 1926 | Committee for Maritime Meteorology favors perpetual equal-quarter calendar |
| A.D. 1927 | League of Nations requests governments to study subject |
| A.D. 1928 | Bills and resolutions in U. S. Senate and House; formal hearings are held |
| A.D. 1929 | Eisenach meeting of non-Roman churches again urges calendar revision. Resolutions introduced in U. S. Congress request President to call international conference on calendar revision |
| A.D. 1930 | The World Calendar Association, Inc. organized in New York |
| A.D. 1931 | Preparatory Committee League of Nations in June; International Conference on Calendar Revision in October |

A.D. 1932    Many international calendar committees formed favoring The World Calendar; Universal Christian Council inaugurates study on the matter

A.D. 1933    President Hoover's Committee on Social Trend urges importance of calendar simplification

A.D. 1934    Protestant Episcopal Church approves fixed Easter in the 12-month equal-quarter World Calendar; governing bodies of American Lutheran and Methodist Episcopal South approve; Congress of Fraternal Insurance Societies endorses World Calendar. United Press questionnaires show that American clergy favors 7 to 1 The World Calendar, and a stabilized Easter 10 to 1. United Press in survey of American business opinion shows a preference for World Calendar to the 13-month plan by about 2½ to 1.

A.D. 1935    National Federation of Belgian Chambers of Commerce favors Easter stabilization and a 12-month calendar revision. Many national scientific associations approve World Calendar. National Academy of Sciences rescinds 1929 resolution which endorsed 13-month calendar. New York State Chamber of Commerce endorses it. American Statistical Association committee on calendar reform after study prefers 12-month equal-quarter plan.

A.D. 1936    Labor Congress of American States at Santiago, Chile, approves perpetual World Calendar; International Labor Organization in resolution mentions World Calendar and requests League of Nations to study question further; London, British and Empire Chambers of Commerce approve; U.S. Central Statistical Board urges revision; Universal Christian Council, after four years' study, approves stabilized Easter based on perpetual 12-month equal-quarter plan

A.D. 1937    National Education Association and World Federation of Education Associations favor a world calendar. League of Nations in March presents Chilean Draft on The World Calendar to State and non-State Members, 14 governments accept in principle; the League in September tables further activities on calendar revision

A.D. 1939    General Federation of Women's Clubs at San Francisco approves resolution endorsing The World Calendar

A.D. 1940    National Federation of Business and Professional Women's Clubs, Council of Bishops of Methodist Church, American Psychological Association approve The World Calendar

A.D. 1941    Milwaukee Society of Accountants approves the new World Calendar

A.D. 1942     Teachers College of Columbia University offers Fellowship on the study of the calendar. The National Conference on Uniform State Laws in its Committee on Scope and Program issues opinion believing The World Calendar most desirable. Lions Clubs in resolution endorse new time-plan

A.D. 1943     Other Lions, Kiwanis and Rotary Clubs endorse The World Calendar. Fraternal Congresses of New York, Maryland, District of Columbia and State of Washington approve. Peoples Mandate Committee for Inter-American Peace and Co-operation passes a resolution commending The World Calendar. Study committees are being formed by National Retail Dry Goods Association, Rotary International, American Institute of Accountants, Illinois Agricultural Association, advertising clubs, etc.

# INDEX

## A

"A.D." and "B.C.," 32-33
Adoption, procedure, 125-127
Advantages of World Calendar, 91-102
Adventists, Seventh-Day, 63
Age of Earth and Man, 8-9
Agriculture, and The World Calendar, 94
Akhenaton, 19
Analogies, three, 86-90
Annual cycle. *See* Year
Approval by 14 Nations, 66
Armelin, M., 54
Armistice Day, 105, 110
Ashbaugh, Lewis E., 70
*Astronomical Kalendar,* 47
*Atlantic Monthly,* 119
August, naming of month, 33
August and July, length of, 33-34

Augustan and Constantinian changes, 32-38
Augustus Caesar, 32-34

## B

"B.C." and "A.D.," 32-33
Bearce, Henry W., 121, 123
Birthdays, 99
Bissextile day, 29, 81
Black, F. A., *The Calendar and Its Reform,* 24
"Blank" days, 80
Books, ref., 10, 12, 18, 19, 20, 21, 24, 27, 28, 32, 33, 34, 35-37, 43, 47, 51, 57, 58, 67, 75-76, 88, 89
Borchardt, 17
Bosco, Sacro, *De anni ratione,* 33
Brady, John, *Clavis Calendaria,* 33-34
Breasted, James Henry, 17, 18, 19
Bridges, Robert, 58

# EXPLANATORY NOTE

Ever since I advocated The World Calendar, it has been my steadfast conviction that the interest of everyone would best be served were the new calendar to be adopted with a minimum of disturbance. Such a time is Saturday, December 30, 1944 when there is a natural blending of the two calendars.

It has been suggested on pages 121-123 that The World Calendar can be adopted on other days and dates: Sunday, July 1, 1945, Friday, March 1, or Wednesday, May 1, 1946. The adoption on the first date would cause the loss of one day, while its adoption on the other suggested dates would cause a loss of two days. Some believe that this adjustment of a lost day would not be too disturbing. However it is my firm opinion that greater harmony would ensue and more general acceptance be assured were The World Calendar to be put into operation without the loss of a day at the end of 1944—when Saturday, December 30, in both calendars, would be followed by December W, a World Holiday, and the next day, Sunday, January 1, 1945 would become the first year of the new time-plan.

I strongly urge its adoption at that time. To delay would bring the next "blending" date to Sunday, January 1, 1950. This needless delay would postpone the many advantages and aids for everyday living, for war and postwar plans, and for the rehabilitation that confronts us.

ELISABETH ACHELIS